'A book that will help us understand the most profound economic changes of our time.' *New York Times*

'In this book I try to tell, in political and human terms, one of the oddest stories in contemporary history: how the world's biggest and most critical industry came to be dominated by seven giant companies: how the Western governments delegated much of the diplomatic function to them; how their control of oil was gradually countered by the producing countries, until in October 1973 it appeared to be suddenly wrested from them. And how since then the seven companies, still the giants of world trade, have found themselves caught on a political tightrope balancing between the demands of their Western oil-consumers and their partnerships with the producers.' From the author's introduction

D0763620

Anthony Sampson is the author of the bestselling
The Sovereign State of ITT and *The Arms Bazzar*, both
available in Coronet.

The Sovereign State of ITT

'An original, topical, sensational, truly significant book.'
Evening Standard

'A lucidly written and very readable account of ITT.
Mr Sampson's book is likely to be widely read.'
Sunday Telegraph

'This book is very, very good – and unquestionably the
best book ever written about ITT.' *New York Times*

The Arms Bazaar

'Brilliant – has the impact of a major scoop.'
New York Times

'Personalities and human drama, international
affairs, relationships and rivalries within and between
huge companies and governments are woven into a
gripping story.' *Daily Telegraph*

'A splendid, absorbing, shocking tale.'
Edward P Morgan, In the Public Interest

The Seven Sisters

The Great Oil Companies and the World They Made

Anthony Sampson

CORONET BOOKS
Hodder and Stoughton

Dedicated to the memory of
Ivan Yates
1926–1975

Copyright © 1975 by Anthony Sampson

First published in Great Britain 1975 by
Hodder and Stoughton Limited

Coronet Edition 1976
Third impression (updated) 1980
Seventh impression 1985

Printed and bound in Great Britain for
Hodder and Stoughton Paperbacks,
a division of Hodder and Stoughton Ltd.,
Mill Road, Dunton Green, Sevenoaks,
Kent (Editorial Office: 47 Bedford
Square, London WC1 3DP) by
Cox & Wyman Ltd., Reading

ISBN 0 340 21323 X

Contents

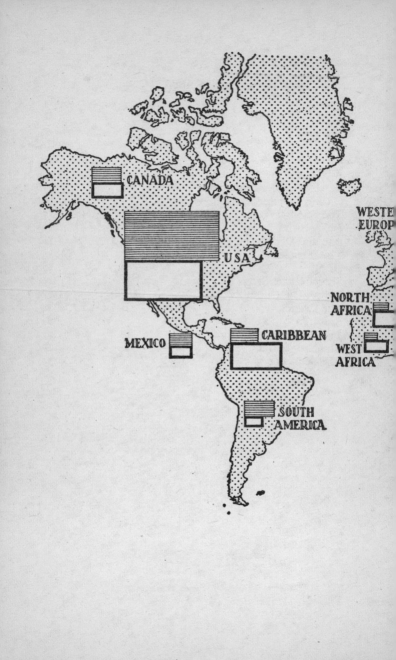

CANADA

USA

MEXICO

CARIBBEAN

SOUTH
AMERICA

WESTE
EUROP

NORTH
AFRICA

WEST
AFRICA

WORLD OIL SUPPLY AND DEMAND

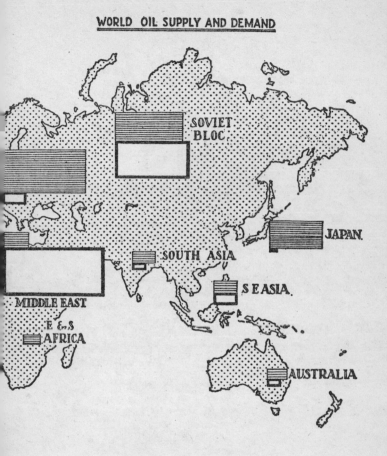

SOVIET BLOC

JAPAN

SOUTH ASIA

S E ASIA

MIDDLE EAST

E & S AFRICA

AUSTRALIA

WORLD OIL
CONSUMPTION
PRODUCTION

Source: BP Statistical Review of the Year (1975)

Introduction

In this book I try to tell, in political and human terms, one of the oddest stories in contemporary history: of how the world's biggest and most critical industry came to be dominated by seven giant companies; how the Western governments delegated much of the diplomatic function to them; how their control of oil was gradually countered by the producing countries, until in October 1973 it appeared suddenly to be wrested from them. And how since then the seven companies, still the giants of world trade, have found themselves caught on a political tightrope, balancing between the demands of their Western oil-consumers and their partnerships with the producers.

I have approached this task not as a specialist or an economist, but as an enquiring journalist with a writer's licence to talk to anyone and to travel anywhere. There is already a vast literature of oil. There are exhaustive company histories allocating praise to all executives. There are technical studies of the economics of the industry. There are romantic narratives of spudding and gushers in the desert. And there are radical attacks on the intrinsic evils of the companies and the cartels. But there has been very little that describes in human terms how these companies grew up, how ordinary men became caught up in extraordinary exploits, and how the Western nations became dependent on these strange corporations.

I had first become interested in the oil companies in the early sixties, while they were competing hectically across Europe, when I wrote a long profile of Shell in my *Anatomy of Britain*, which brought me into contact with oilmen, including Enrico Mattei, the Italian tycoon who first popularised the phrase 'The Seven Sisters' Thereafter I became interested in the problem of the control of multinationals, which I wrote about in my book *Anatomy of Europe* (*The New Europeans*), and later in the *Sovereign State of ITT*. My interest in the oil companies, as the oldest and biggest of the multinationals, was greatly stimulated in Washington, where I was spending a year from September 1973, coinciding with the embargo and the energy crisis.

The role of big companies in influencing foreign policy and changing global relationships is an engrossing subject for study; but it is difficult to describe or analyse the workings of the companies without a body of reliable documentary evidence. It was not until the crisis of 1973 that a great deal of such documentation came to light; particularly through the exhaustive investigation by the Multinationals Subcommittee of the Foreign Relations Committee, under Senator Church. The detailed testimony, the subpoenaed memos, secret agreements and cable traffic that emerged from those hearings, some published, some still unpublished, provide a unique glimpse into the methods and workings of the men inside the oil companies. To this copious source this book is much indebted.

With this basis of reliable documentation, I have tried to trace the narrative in an informal style, portraying events and decisions through the eyes and minds of the people that were involved. There are many economists and oilmen who are inclined to depict their industry as subject to iron laws of supply and demand which allow little scope for human choice or initiative: they suggest that if there is any real master of the business, it is the slippery fluid itself, which has changed the balance of the world. Certainly the dark, greasy character of crude oil plays an important part in this story. But looking at the key turning points in oil history, it is difficult, I believe, to ignore the decisive roles of a handful of masterful figures who imposed their personalities on the business; from Rockefeller at the beginning, through Deterding and Teagle between the wars, to the post-war intruders like Getty and Mattei, to the new oil tycoons of our time, including the Shah of Iran and Sheikh Zaki Yamani. It is through the views and attitudes of such men that I try to convey part of this story.

This book does not set out to explain all the intricacies of the oil business and the energy crisis; it does not touch on natural gas, coal or nuclear fuel. Nor does it pretend to give a comprehensive picture of Middle East politics, and the Arab–Israel question which are exhaustively dealt with elsewhere; the fact that Arab nations here receive much more space than Israel reflects only the fact that they have the oil, and the close links with companies which are the subject of the book. It is the chief aim of the book to convey the history of oil through the people involved in it, and to show the attitudes and the psychological changes that have lain behind the extraordinary revolution of our times.

In writing this book, particularly in the early chapters, I owe much of course to the many historical analyses of the oil business, and I have made much use of the standard works which are acknowledged at the end of each chapter. But I have supplemented these sources with scores of conversations with many of the chief participants in the events I describe, who have helped to give the atmosphere and attitudes behind the decisions. I have travelled as widely as possible to the chief centres of the oil industry: to the headquarters of the companies, in London, New York, San Francisco, Pittsburgh; to the historic oilfields in Pennsylvania and Texas; to meetings of OPEC in Vienna and Algiers; and to the chief oil-producing countries in the Middle East. By quoting views and assessments from all sides—from the companies, from the Western governments, from the producing governments, and from the critics of the industry—I have tried to present a fair balance to my narrative, and to let the reader decide for himself where if anywhere the blame lies before giving my own conclusions in the last chapter.

This is essentially a book about the political consequences of oil, not its economic basis or its engineering achievements, remarkable as they are. I make no apologies for trying to simplify the problems into the basic political questions; for the more one looks at the critical decisions—whether in 1954, 1960 or 1973—the more it appears from the testimony of the decision-makers, that they were taken on the basis of relatively simple choices. Behind all the intricacies of specific gravities, premiums, discounts and buy-backs, many of the problems of oil have been crude, in both senses.

I have tried to avoid jargon as far as possible, to let the narrative flow easily. The names of the companies present a special difficulty, for some of them have changed in the course of their long lifetimes; BP was originally Anglo-Persian, then Anglo-Iranian; Exxon was originally Standard Oil of New Jersey, then known as Esso, as it still is in Europe; Mobil was originally two companies, Standard Oil of New York (Socony) and Vacuum, then Socony-Vacuum, then Socony-Mobil. In each case I have decided to call the companies throughout by their current names—BP, Exxon, Mobil. While their names have changed, their character has been a continuous development.

My debt to individuals is very great, for a book of this kind

depends heavily on the co-operation and helpfulness of the participants in the story. Within the companies themselves I have been
received with varying cordiality, but only Texaco have been
downright unco-operative (for reasons which may emerge in the
book). Shell have been the most helpful and open, though touchy
(readers may note that they alone refused to allow their symbol to
appear alongside the others on the jacket). Mobil have been the
most talkative, and BP the most uniformly loyal.

Many of my informants inside the companies would prefer not
to be mentioned; but many executives and former executives went
to great trouble to put their viewpoints, in spite of my sceptical
attitude. In Exxon, I am specially grateful to Ken Jamieson,
George Piercy, Howard Page, Steve Stamas, Emilio Collado and
Bill Slick. In Mobil, I enjoyed outspoken talks with Rawleigh
Warner the chairman, Bill Tavoulareas the president, Herb
Shmertz and Andrew Ensor. In Socal, Dennis Bonney and
George Ballou went to special trouble to convert me. In Gulf, I
owed much to Jimmy Lee the president and Paul Sheldon, as
well as to several former employees. About Texaco, I resorted
more to former company men and shareholders. For recollections
about Texaco's early history, I am grateful to Andrew Jackson
Wray, the son-in-law of the founder, for his exuberant hospitality.

Within BP, I am grateful to Sir Eric Drake the chairman, to
Peter Walters, John Sutcliffe, Julius Edwards and Dr. Ronald
Ferrier, the company historian, for interesting arguments; and to
former employees including Geoffrey Keating, Martin Jay and
Archibald Chisholm, whose information about the early history of
Kuwait has been invaluable, and who has kindly shown me proofs
of his important forthcoming book: *The First Kuwait Oil Concession Agreement* (London 1975). Within Shell, I have had helpful
discussions with Gerry Wagner the chairman, Sir Frank McFadzean the vice-chairman, and two previous chairmen, Sir David
Barran and John Loudon; while in Houston I was helped by the
president of U.S. Shell, Harry Bridges. I am also grateful to
André Bénard, Dan Samuel, Geoffrey Chandler and Ian Skeat,
who have all tried to inform me without trying to brainwash me.

Among many officials of Aramco, I am grateful to Joseph
Johnstone and Ted Phillips of the New York office, to Mike
Ameen in Washington, and to Frank Jungers, the president in
Dhahran. To all these company men I am grateful for their time
in trying to convince me of their company's viewpoint. But I am

not indebted in any other sense; all my travelling and the costs of research have been at my own expense.

I have gained much from visits to Houston, Texas, and from the hospitality of my brother-in-law Dr. Philip Bentlif. I have learnt many insights from Houstonians, particularly from John Jacobs, Philip David, and Manro Oberwetter; James Clark kindly gave me advice and the use of his remarkable library of oil books. Elsewhere I have been helped by many independent oil experts and critics with their sidelights on the big companies, including Paul Getty, Henry Schuler, Charles Perlitz, George Spencer of *Oil Week*, and I am especially grateful to my friend Thomas Mullins in London, for his shrewd observations.

Outside the companies, I am indebted to many politicians and diplomats who have given their side of the picture, and of past decisions. In the United States they included John McCloy, George McGhee, James Akins, Abe Fortas, George Ball, Thomas Kauper, Senators Edmund Muskie, Jacob Javits and Abraham Ribicoff, and Congressman Les Aspin. In the Federal Energy Administration I received useful advice from Mel Conant and Pamela Kucser. I owe a very special debt to Senator Church, who encouraged me to pursue my enquiry, and to my friend Jerome Levinson, his chief counsel on the Multinationals Subcommittee, who helped patiently to guide me through the maze of evidence that he had assembled; and to others of the sub-committee staff, John Henry, Jack Blum and Bill Lane, who gave much of their time. In Britain, I have had useful advice from many past and present diplomats and politicians concerned with oil, including Sir Kenneth Younger, Sir Denis Wright, Sir Harold Beeley, Lord Balogh, Eric Varley, Hugh Fraser and several serving ambassadors. At Chatham House, the observations of Louis Turner have been always stimulating.

From the OPEC side, I have enjoyed talks with three earlier secretaries-general: Fuad Rouhani in Iran, Nadim Pachachi in London, and Abderrahman Khene in Vienna; and also with the current incumbent, Chief Feyide. I also had a useful conversation with the Secretary-General of OAPEC, Ali Atiga in Kuwait. In St. Moritz I talked with the Shah of Iran, and in Teheran and Vienna with his oil minister, Dr. Jamshid Amouzegar. In Riyadh I talked with Sheikh Zaki Yamani the oil minister, with Dr. Abdulhadi Taher the head of Petromin, with Hisham Nazer the Minister of Planning, and many others. In Bahrain I talked to

Yusuf Shirawa the oil minister, and Ben Mubarak the foreign minister. Among Kuwaitis I talked with Abdul Rahman Atiqi and Ali Khalifa of the oil ministry, Ahmad Jaafar and Suleiman Mutawa of the Kuwait Oil Company, and Abdullatif Al-Hamad of the Kuwait Fund: I was also grateful in Kuwait to Mrs. Violet Dickson, who allowed me to record her recollections of the first negotiations for the oil concession. Among diplomats, I have specially appreciated the help of Ambassador Brahimi of Algeria in London and Ambassador Zahedi of Iran in Washington.

Among journalists in Washington I have owed much to Joseph Kraft and to Morton Mintz and Laurence Stern of the *Washington Post*. In London I have been much helped by Patrick Seale, Michael Addams, Robert Stephens of *The Observer*, Richard Johns of the *Financial Times*, Peter Hillmore of *The Guardian*, and Mario Ciriello of *La Stampa*. Among oil specialists I have had useful advice in New York from Walter Levy and from Wanda Jablonski of *Petroleum Intelligence Weekly*: and in Beirut from Fuad Itayim, the owner of *Middle East Economic Survey*, and his expert news editor, Ian Seymour, who has been generous with his assistance.

These varied sources have, I hope, helped to convey the story from a range of different viewpoints which I have tried to depict accurately and fairly. But I have not hesitated to express my own views, particularly in the last chapter which represents only my own opinions.

My great indebtedness to written sources will appear in the notes to the book. In trying to explain the economics of the industry, I have tried to navigate between the rival sirens among the experts, notably Paul Frankel, whose *Essentials of Petroleum* in 1946 pioneered the science of oil economics; Edith Penrose, whose book *The International Petroleum Industry* (1968) analyses the financial structure and functions of the seven sisters; and Maurice Adelman, whose *World Petroleum Market* (1972) confutes many earlier beliefs about competition and cartels. To the first two writers I am specially grateful for personal advice and suggestions.

Throughout I quote repeatedly from the hearings of Senator Church's Multinationals Subcommittee, referred to as *Multinational Report*, which includes an important history of the relations in 1974 and 1975 as Parts 4–9 of the series on Multinational Corporations and U.S. Foreign Policy. (Parts 8 and 9 had not been published in volume form by the time of going to press, so are quoted without page references.) The report of the Subcommittee, published in January 1975, referred to as *Multinational Report*, includes an important history of the relations between the U.S. government and the companies, based on the evidence from the hearings.

In completing the book, I owe a great deal to the help of my two editors. Alan Williams of Viking in New York provided shrewd suggestions and comments which much enlivened my task. Alan Gordon Walker of Hodders in London helped to solve structural problems with steady patience and humour. For consistent support I could rely on my literary agents, Michael Sissons of A. D. Peters in London and Sterling Lord in New York. For organising the whole operation, from travel to research to typing to checking, I again had the indispensable help of my assistant, Alexa Wilson. And for research, advice, and encouragement throughout the project I have been kept going by my wife, Sally.

June 18, 1975

In this new edition I have added an additional chapter describing the new oil crisis of 1979 which is really a continuation of the story in earlier chapters, and which adds to earlier warnings, bringing the story of OPEC and the relations with the Seven Sisters up to date.

July 3, 1979

1

Who Controls?

Oil friendships are greasy.

Calouste Gulbenkian

I was watching the arrival of delegates, at the Palace of Nations outside Algiers in March 1975, feeling as dazzled and bewildered as if it were the climax of a long opera. Lit by the bright African sun, they appeared under the triumphal arch with a pomp and fervour which suggested either a victory march or a revivalist rally. They strode through in teams, past the Algerian guard of honour in white turbans and green cloaks, anxiously fingering their swords, while the watching officials gave each team a round of applause. The Sheikh of Abu Dhabi, with his immaculate black spade beard, swirled his dazzling white robes, followed by a group of shrouded white shapes, with only high-heeled shoes to distinguish them. The tiny President of Ecuador, General Lara, bright with medal-ribbons and gold lace, stepped briskly between his towering aides. The Sheikh of Kuwait beamed underneath his wide moustache, followed by a guard in bright scarlet uniforms. The Libyan revolutionary Major Jalloud loped through, his lean shape enveloped in a flapping double-breasted suit. The President of Venezuela, Carlos Andres Perez, with his sideburns almost meeting his mouth, processed solemnly past, accompanied by much clapping from his waiting publicity-men. The strong man of Iraq, Saddam Hussein, strode in front of a group of stony-faced aides. And between the Arab teams walked the urbane figure of the Shah of Iran, chatting casually to his ministers, like a well-groomed banker joining this strange gathering of rulers and revolutionaries.

Inside the great circular hall, under the white dome, the gathering seemed even more like a religious rally. Round the circle the delegations sat in their sectors, in alternating colours: first a sector of black curly hair and dark suits, then a mass of bright white burnouses and circlets, like a sea of table-napkins. Up on the rostrum stood the President of the host-country, Houari Boumédienne, making one of his famous marathon speeches, with a combination of Cartesian logic and Arab rhetoric, interspersing his arguments with the phrase: 'Your Majesties, Brothers'. Above the rostrum hung a mystic symbol of four white incomplete circles on a blue background, which could only with difficulty be deciphered as the magic word: OPEC. This was indeed the first summit meeting of the Organisation of Petroleum Exporting Countries.

Few people eighteen months before could have imagined that such a gathering could be brought about through a common interest, not in religion or arms, but in a greasy black fluid. Even knowing the cause, it was difficult to remember in the flood of grandiose rhetoric. Oil, the Shah explained (once again), was a noble fluid which must be properly conserved. The conference, boomed the President of Venezuela, was not merely concerned with ordinary energy, but with 'another kind of energy which is rarer; moral energy'. Oil, said Major Jalloud, must be the foundations of a new economic order. It must, said the Shah, be the means of making 'a world in which it will be an honour for us to live'. And President Boumédienne reiterated his favourite theme: oil was not merely a fuel, but the source of life itself. The oil-producing countries must be the spearhead of the revival of the Third World, and the means to an equitable new system of world justice.

As I watched the conference, I was thinking not so much about the potentates, as about the unpredictable fluid which in a year had apparently turned the world upside down. Certainly the arrangements of the conference suggested some kind of revolution. While the delegations from the Middle East or Latin America marched in and out, driven off in their shining black Citroens to banquets and private meetings, the Western diplomats and journalists came in by the side doors, milling around the crowded convex spaces outside the circumference, trying to pick up news of what Venezuela had said to Algeria, or what Saudi Arabia thought about Libya. Or they sat inside the hall, listening through earphones to the monotone translations of the stern speeches: arraigning the 'manoeuvres and manipulations of certain powers', emphasising that 'countries living beyond their means must accept the transformation that is inevitable'. The speakers were not extravagant or even bitter, in the older tradition of Third World conferences. They sounded responsible and even patronising, conscious (in the words of Boumédienne) of the 'onerous privilege' of the producers of oil.

Oil, which had sprung so many surprises on the world in the past century, had now apparently achieved its master-trick; with a turn of the circular stage, a whole new scene had come into view, with another complete cast of characters. Watching this exotic pageant I found myself marvelling that the West should ever have become dependent on such an unreliable commodity. In its murky trail, from the hills of Pennsylvania to the flatlands of Texas to the oil compounds of Saudi Arabia, the black stuff had always seemed to be spurting up in the most impossible places, one moment in excessive quantities, the next moment threatening a terrifying shortage, as if to exasperate its millions of dependents. Nor was it, like most other commodities, something which consumers could effectively hoard or stockpile to protect them against its quirks. The only economic reservoir was under the ground in its original oilfields, where it was subject to the control of the people who lived above it. It was like water in its bulk and its problems of storage, but unlike water it had to be carried across the world, to transform the economies of nations.

It all seemed very confident and cheerful, this brave new world. Oil seemed to have united these disparate countries, not only to keep up its price, but to dissolve their political differences. The Majesties sat next to the Brothers, the Africans congratulated the Arabs; and at the close of the conference President Boumédienne suddenly presented himself as the peacemaker between two of the bitterest enemies in the Middle East who for years had been raiding across each other's frontiers. Saddam Hussein of Iraq walked across the hall to the Shah of Iran. The delegates all round the horseshoe stood up and clapped, and the two leaders kissed and embraced.

It was all done by oil: by an oil cartel. Behind all the idealism about the Third World, the delegates knew quite well that they could not achieve anything without a high price for the commodity on which they all depended. President Boumédienne roundly condemned 'the practice of setting up exclusive clubs or "in-groups" .' But that was precisely what OPEC was intended to be: in the words of one of its founders, Perez Alfonso: 'we have formed a very exclusive club'. Unless that inner club held together they had little hope of helping anyone outside it.

On the surface the unity was magnificent; but there were many hints of dissension. Among the final speeches there were significant omissions. There was not a word from the Sheikh of Kuwait, not a syllable from the other Sheikhdoms, from Qatar, Abu Dhabi or Dubai. And most significant, no reaction at all from Sheikh Zaki Yamani, the oil minister from Saudi Arabia, the biggest producer of all. And in the corridors, between the brave speeches, delegates seemed less confident, as if there were still a spectre at the feast. Everyone knew that the Saudis, with their vast oil production, held the key to the cartel. But the Saudis made no bones of their scepticism about mixing up oil with the politics of the Third World.

Why had the demand for oil from Abu Dhabi dropped so precipitately? Why were so few tankers now arriving at Kuwait? Who exactly was deciding to take less oil from Iran?

And there was now a very visible coldness between Yamani and the second most important exporter of oil, the Shah of Iran. In the previous month I had talked with both of them, and each had explained with some asperity his suspicions about the other. The Shah was convinced that 'that fellow Yamani' was taking his policy from the Americans, agreeing to reduce the price of oil in return for special favours to the Saudis. The Shah was determined that the price of oil should go up still further; for the sake of the West (he hastened to explain) as much as for the producers. But the Saudis regarded the Shah as dangerously callous to the world's economy, and of being obsessed by his short-term need for money, for arms and self-aggrandisement. Now, at Algiers, the disagreement between the two was publicly evident, and the argument about price was still not settled. And it was the price that was critical to every cartel. Yamani had insisted to me that this cartel was different from all others; because the stronger members, led by the Saudis, had no need to undercut the weaker members: having got the right price they were prepared to cut back indefinitely to maintain it. Would this be a new phenomenon in world history, a strong cartel controlled by sovereign states, by producers rather than consumers? Or did that make it all the more fragile? Would the old powers of the companies and the consumers reassert their role, to divide and to rule? Would the circular stage turn again?

THE SISTERS

There were no men from the oil companies at Algiers, and the names of Exxon or Shell were hardly mentioned, as if they belonged to a deposed regime. But in New York or London, the company experts were carefully watching and analysing the news and the hints that emerged. They were calculating the longer credits, the lower premiums, the increasing discounts, redirecting their tankers in mid-ocean from one loading port to another, to take cheaper oil, watching for each sign of a break in oil prices. Was the solid front dissolving at the edges, as the supply went up and the consumption went down? Were the palmy days of the 'sixties returning, when Iran and Saudi Arabia were desperately competing, not to demand more revenue for their oil, but to produce more? Would the whole OPEC solidarity gradually fall apart provided that their Western enemies did not crow about it? Were the oil companies, provided they played their cards cleverly, still able to regain their command, because they controlled the markets on which every oil producer depended?

Eighteen months earlier the West had taken for granted that the world's oil was in the hands of the oil companies, and in particular of the seven giant corporations known as the Seven Sisters whose names were proclaimed on the big signs that stick up like lollipops from the roadsides in countries all over the world: Exxon (or Esso), Shell, BP, Gulf, Texaco, Mobil and Socal (or Chevron). It was these companies that had built up the international oil industry over the past fifty years, growing up into some of the biggest corporations in the Western world. It was they, not the countries, who were accused by their critics of being a cartel.

And still now, it was hard to believe that they had lost their power and influence. With their complexity, range and resources, they were institutions that had appeared to be a part of world government. Their executives could fly between Pittsburgh and Kuwait, between San Francisco and Saudi Arabia, as casually as across their own state. Their computers could analyse the supplies and demands of half the countries of the world. Their boards could allocate hundreds of millions to bring into being a new oilfield, a new harbour, or a new trade-route. Each of the seven had now lasted more than fifty years, longer than many of their nation-clients. Their skyscraper headquarters, sticking up from their domestic surroundings, seemed to evoke a new world where nations themselves were obsolescent.

For decades the Companies (with a capital C) seemed possessed of a special mystique, both to the producing and consuming countries. Their supranational expertise was beyond the ability of national governments. Their incomes were greater than those of most countries where they operated, their fleets of tankers had more tonnage than any navy, they owned and administered whole cities in the desert. In dealing with oil they were virtually self-sufficient, invulnerable to the laws of supply and demand, and to the vagaries of the stock markets, controlling all the functions of their business and selling oil from one subsidiary to another. Shell oil was pumped from Shell oilfields into Shell tankers, onto Shell refineries into Shell storage tanks, and through Shell pipelines to Shell filling-stations. They were the first of the global giants ('Exxon was a multinational corporation', said their 1973 Annual Report 'at least fifty years before that term was commonly used'). And they had expanded into other industries, including petro-chemicals, coal-mining, nuclear power, to make not only energy but plastics, fertilisers, and drugs.

But it was oil, with its huge geographical spread and political consequences, which had been the source of their unique influence, making them the masters of half the world's trade and, most remarkable of all, giving them control over concessions whereby nations sub-contracted part of their sovereignty. In the Middle East there were two kinds of maps: some showing the names and outlines of the nations, most of them comparatively new; and others showing the region cut up into squares along the coast, marked with initials—IPC, KOC, ARAMCO, AOC—representing the consortia of oil companies, nearly always including some of the seven sisters. To the companies it was these squares which were the real geography: Saudi Arabia was Aramco-land; Iran meant all the Seven; Kuwait was Gulf and BP. To the Arab politicians the companies had loomed larger than the Western nations themselves, for it was they who could decide how to allocate economic growth between one state and another, and who impressed their own characters on the emerging nations.

Travelling between the skyscrapers of the sisters, I had been conscious of moving within the same separate world, though seen from different longitudes. In Pittsburgh the solid stone monument of Gulf, with a pyramid on the top, has dominated the city for forty years, still linked with the wealth of the first Pittsburgh family, the Mellons. In San Francisco the tower of the Standard Oil Company of California, or Socal, looms over the bay like a fortress, the richest company in the city. In New York two of the sisters face each other across 42nd Street, but with almost opposite viewpoints; Texaco, inside the Chrysler building, cultivates a reputation for meanness and secrecy, while Mobil over the road is the most loquacious and extrovert of them all, churning out explanations, complaints and counterattacks through the TV channels and newspapers. Across the Atlantic in London, the headquarters of BP rises up from its own piazza, announcing with its name Britannic House that here is an oil company that is part of the nation's patriotism, half-owned by the government.

Within these corporate citadels there was little doubt that they were competing ferociously with each other. Their inhabitants seem to belong not so much to the world of 'oildom' as of Mobildom or Gulfdom, and with some coaxing they could be persuaded, I found, to criticise each other. Yet each had been linked to all the others through a web of joint ventures and concessions across the globe, from Alaska to Kuwait: sharing now with one partner, now with another, in different permutations. It was this strange cavorting of the sisters, competing one moment and conniving the next, which had made them such an enduring subject of suspicion and investigation by politicians, economists and nationalist leaders.

The band of sisters had been led by two giants, Exxon and Shell, who for the past sixty years had been the prototypes of the sophisticated international company. Their rivalry across the continents had been a long sub-plot to modern history, financing whole nations, fuelling wars, developing deserts. Their commercial ambitions were fraught with diplomatic consequences: the revolutions in Iraq, the separatist movement in Scotland or the civil war in Nigeria. They had seemed often enough like private governments, to which the Western nations had deliberately abdicated part of their diplomacy, and sometimes appeared as if they were Atlas himself, bearing the world on their shoulders. Inside their corporate headquarters, which had seen so many wars and crises come and go, it was hard to believe that the new phenomenon of OPEC was anything more than another temporary setback in the long triumphant progress of the oil companies. For they represented much more than themselves: they were a central part of the whole economic system of the West.

EXXON

In the middle of Manhattan, in the line of cliffs adjoining the Rockefeller Center, is the headquarters of the most famous and long-lived of them all: the company known in America as Exxon, and elsewhere as Esso, and for most of its hundred years' existence as Standard Oil of New Jersey or simply Standard Oil. It is a company which perhaps more than any other transformed the world in which we live. For much of its life it was automatically associated with the name of Rockefeller, and some links still remain. The family still own two percent of the stock; Nelson Rockefeller once worked for it in Venezuela; and the desk of the founder, John D. Rockefeller I, is still preserved as a showpiece at the top of the building. But Exxon has long ago outgrown the control of a single family. It is, by assets, the biggest company in the world. It has 300,000 shareholders, its subsidiaries operate in a hundred countries, and in 1973 its profits were a world record for any company in history: $2·5 billion.

The tranquil style of Exxon's international headquarters seems to have little in common with the passionate rhetoric of Arab politicians in Algiers. Beside a bubbling fountain and pool on Sixth Avenue, the fluted stone ribs soar up sheer for fifty-three storeys, and inside the high entrance hall is hung with moons and stars. On the twenty-fourth floor is the mechanical brain of the company, where the movements of its vast cargoes are recorded. A row of TV screens are linked with two giant computers, and with other terminals in Houston, London and Tokyo, in a system proudly named LOGICS (Logistics Information and Communications Systems). They record the movement of five hundred Exxon ships from 115 loading ports to 270 destinations, carrying 160 different kinds of Exxon oil between sixty-five countries. It is an uncanny process to watch: a girl taps out a question on the keyboard, and the answer comes back in little green letters on the screen, with the names of ships, dates and destinations across the world. From the peace of the twenty-fourth floor it seems like playing God—a perfectly rational and omniscient god, surveying the world as a single market.

Up on the fifty-first floor, where the directors are found, the atmosphere is still more rarefied. The visitor enters a high two-storey lobby with a balcony looking down on high tapestries; the wide corridors are decorated with Middle East artifacts, Persian carpets, palms, or a Coptic engraving. It is padded and silent except for a faint hum of air-conditioning, and the directors' offices are like fastidious drawing rooms, looking down on the vulgar bustle of Sixth Avenue. It all seems appropriate to Exxon's reputation as a 'United Nations of Oil'.

But in this elegant setting, the directors themselves are something of an anticlimax. They are clearly not diplomats, or strategists, or statesmen; they are chemical engineers from Texas, preoccupied with what they call 'the Exxon incentive'. Their route to the top has been through the 'Texas pipeline'—up through the technical universities, the refineries and tank farms. The Exxon Academy, as they call it, is not a university or a business school, but the giant refinery at Baton Rouge, Louisiana, Watching the Exxon board at their annual meeting, I found it hard to imagine them as representatives of a world assembly. It was true that there were, in 1974, two foreign directors— Prince Colonna, the former commissioner of the Common Market, and Otto Wolff, the German industrialist; and there was also one director, Emilio Collado, with experience of government. But the core of the board was made up of the engineers, enclosed in their own specialised discipline.

Ken Jamieson, the chairman and chief executive, a tall cliff of a man with a wide deadpan mouth, was brought up a Canadian, in Medicine Hat, the son of a Mountie; but Texas has since become his adopted home, and he will soon retire there. Jim Garvin the president, an engineer from Virginia, worked his way up through chemicals and Texas, insulated from the world outside oil: he is expected to be chief executive for eight years. Mike Wright, the Chief Executive of Exxon, U.S.A., began as a roughneck in Oklahoma before he, too, went down to Texas, where he now lives: a wiry outspoken champion of the fight for free enterprise. Tom Barrow, a short thick-set man, comes from an old Texan oil family with a shareholding in Exxon: he studied geology in Texas and California, made his name in exploration, and came back to Texas as President of Exxon, U.S.A.; he is likely to be Garvin's successor. George Piercy, the Middle East negotiator, is the director who is most constantly concerned with diplomacy and foreign countries: he has a combative look, with a quiff of hair and a bulldog face, as if built to battle over barrels. He too came up through the pipeline: he became a chemical engineer in Minnesota, and graduated at the Bayway Refinery.

Within their own citadel these men seem confident enough, with some reason: they are directors of a company that has survived for a century, they have acquired great expertise, and they each earn over $200,000 a year. They move in a world enclosed by the rules of Exxon, which belongs to them. 'I think of it as a proprietory relationship,' said Garvin in 1973. 'Like running a company of which I am the owner. It is not just my duty, but my deep personal desire, to keep it in the best shape possible for the men who will come after me.'[1] But once outside their own territory, their confidence easily evaporates. Confronting their shareholders they seem thoroughly nervous, sitting in a row, their fingers fidgeting and their cheekbones working, as they listen to questions about Exxon's African policy, Exxon's salary policy, Exxon's kidnap policy, Exxon's Middle East policy. They know well enough that their company, while one of the oldest, has also been the most hated.

It is in Texas, not New York, that the Exxon men feel more thoroughly at home: and it is the Exxon skyscraper in Houston, the headquarters of Exxon, U.S.A., which seems to house the soul of the company. At the top is the Houston Petroleum Club, with two entire storeys making up a single room, where the oilmen can lunch off steak and strawberries every day of the year. They like to show visitors the view, of which they are justly proud. The flatlands stretch in every direction, broken only by the jagged man-made objects: the domes and tower-blocks in place of cliffs and hills; the curving freeways instead of rivers; the giant road-signs instead of trees. The glaring gasoline signs stick up from the desolate landscape, like symbols leading to some distant shrine: Exxon, Texaco, Shell, Gulf, Exxon. The fluid which has wrought all these changes is concealed from the view: around Houston, there are only a few little pumps nodding in the fields, a few piles of pipelines, to indicate the underground riches. But no-one needs reminding: it was all done by oil.

SHELL

Across the Atlantic another oil city presents itself as a world organization, with a style more lordly and sedate. On the South Bank of London, looking over the Thames, is the great Shell Centre with stone façades and a solid tower, the first of London's skyscrapers, dwarfing the tracery and pinnacles of Parliament over the river. Behind the high walls is the complete city of Shell, with its own swimming-pool, canteens, rifle-range, clubs, reception rooms, cinema and underground passages, as if prepared for a siege. The decor suggests an obsessive introversion, as if devised by a crazy conchologist: shells on the glass doors, shells along the façade, an exhibition of sea-shells in the lobby and a high sculpture in the courtyard which turns out to be made up of shells. Shell men nowadays like to apologise for the building's pomposity, but it may have some symbolism; for it suggests not only the self-containment, but the overwhelming importance of Shell in the economy of the home country—an importance even greater in the other half of its home, in The Hague.

Shell men regard Exxon as a provincial company and they explain how their formative experience has been the opposite. For Shell, as we will see, began not as a domestic oil-producer but as a trader and importer of oil to its home countries. 'They grew up as a Pennsylvania company,' explained a Shell director: 'but we've been in politics all our lives.' Shell's executives have become recognised as a kind of junior diplomatic service. In Britain their traditional nurseries have been Oxford and Cambridge, their children are educated free at public schools, and many of the directors become commanders or knights of the British Empire. Shell men like to cultivate almost the opposite style to Exxon's: while the men in Sixth Avenue prefer not to talk about diplomacy. Shell men prefer not to talk about anything as squalid as profits.

'The Group', as Shell calls itself, has a hundred different faces. In Britain, most people imagine Shell to be purely British. In Holland, they regard Royal Dutch/Shell (as it is properly called) as predominantly Dutch, which it is. In the United States, where it does a third of its business, Americans assume it to be an all-American company with European subsidiaries. It is certainly very cosmopolitan: its top management is balanced between Dutch and English, sixty-to-forty; the shareholders of the whole group are 39 percent British, 19 percent American, 18 percent Dutch; and Shell has tactfully internationalised its staff and subsidiaries. The chairman of the eight men who rule Shell is a Dutch lawyer, Gerry Wagner; the vice-chairman is a combative Scots economist, Sir Frank McFadzean; and there is even a Frenchman, André Bénard, on the board.

'More than any other company,' said a previous chairman, Sir David Barren, in 1967, 'we are so international that nothing can happen in any part of the world without it affecting our interests.' Shell maintains a lofty and sceptical attitude towards governments, and McFadzean sees multinational corporations like Shell as the victims of nationalism and irrationalism throughout the world. 'They are prisoners of their past investments [he wrote recently, dissenting from my own picture of multinationals[2]], even the most puny government can nationalise, and the only redress is to seek compensation.'

The Shellocracy live and work in grander style than their government equivalents, enclosed in their world of private dining rooms, planes and chauffeurs (as they complain about public wastage of fuel, their chauffeurs are circling the block). Looking down on Whitehall from their stone tower, they are above petty national politics, trying vainly to educate politicians in the eternal verities of oil. Sir Frank earns three times as much as the prime minister ($200,000 a year) and other companies look with envy at Shell's elaborate tax-free arrangements.

The self-image of Shell suited well enough with the European mood of the 'sixties, and the glamorisation of big business. It was romantically reflected by the British television series *The Trouble-shooters*, which depicted an international oil company called Mogul, run by a heroic triumvirate of tycoons out-manoeuvring rivals and overcoming disasters, with cunning and the traditional stiff upper lip. The Sheikhs were unscrupulous, the continentals were womanising, the Americans were greedy and naively anti-Communist, and the British diplomats were stuffy and ignorant, getting in the way of the great Company.

But this was not how it looked to a new generation, particularly after the 1968 wave of student radicalism. Shell bravely tried to come to terms with that revolt, as I observed at some seminars between senior Shell managers and radical students, who were invited out to the Shell Country Club to explain their opposition. It was a painful confrontation. The Shell men had come back from the far outposts of the Company, from Venezuela, Indonesia or Brazil, and looked for recognition and acclaim. They saw themselves as a dedicated profession, like doctors; they were not in a separate jungle of business, but an indispensable part of the world community. They had no sinister power, indeed very little power at all in the face of nationalist forces abroad and a watchful government

at home. But the students remained unconvinced. They saw the
Shell men, not as heroic Troubleshooters, nor even as exciting
villains, but as dull, decent men who were trapped in a giant
system of exploitation, who had wrecked the ecology at both ends
of their business, and who were by-passing democracy by their
own autocratic decisions. There was little meeting of minds
between the two sides: 'I don't hear you,' a Scots Shell man kept
sadly repeating, 'I don't hear you.' The oilmen were baffled and
hurt, but basically undeterred: they knew quite well that none of
the students knew anything about liquefaction plants, gathering-
stations or cracking techniques, and the whole massive technology
on which the whole world, including the students, depended.

THE BROKEN BUFFER

The oil companies led by Exxon and Shell, have had good
reason to regard themselves as indispensable to both the West,
and to the oil-producers. It was not just that they were the masters
of engineering and unprecedented capital resources, who even
now were exploring new depths of the North Sea and new reserves in the Arctic, far away from the intrusions of OPEC. It was
also that they had in the past half-century acquired a very special
diplomatic role, and learnt a strategy which had defeated many
earlier threats. Through nationalisations in Mexico or Iran, revolution in Iraq, wars in the Middle East, they had successfully
maintained the steady flow of oil, increasing never too fast to make
the market collapse, never too slow to make consumers go short. If
anyone was expert in the subject of controlling supplies, it was not
OPEC, but they. They still had the control of the world's markets,
and their diplomatic significance rested on the fact that the Western governments had delegated to them, by accident or design,
most of the task of ensuring the supply of the most critical commodity. The companies, they frequently explained, provided a
'buffer' between the producing and consuming governments; or
'a thin lubricating film' in the words of Sir David Barran of
Shell. Or in the words of one oil authority, J. E. Hartshorn, they
do 'the business in between', business which needs 'an international viewpoint detached from the special interest of any one
nation'.[3]

But as the tensions increased between the West and the Middle
East in the early 'seventies, the two were set on a collision course
and the workings of the buffer looked increasingly doubtful. The

new powers of OPEC had been established in specific opposition, not to the Western countries, but to the companies; and while the leverage of OPEC increased with the prospect of shortage, the leverage of the companies became much less certain. The final collision came in Vienna on the night of October 11, 1973: an event which was scarcely noticed in the world's press at the time, but which must rank as one of the most critical encounters in the history of oil.

The oil company men had gathered in Vienna four days earlier, led by George Piercy of Exxon, to negotiate with OPEC for a new price of oil, and they had been briefed in London beforehand to offer no more than twenty-five percent more. But they arrived in the midst of a political turmoil, for the Middle East War had just broken out. The more militant Arabs were enraged by the American support for Israel, and were exchanging information and details from Washington. The war had added fervour and unity to their bargaining, but they were still negotiating. The Iranians wanted the price to go up from $3 to $5 a barrel, and Sheikh Yamani from Saudi Arabia, who led the OPEC team, began by asking for a doubling of the price. The oilmen were adamant, and in the meantime the political atmosphere was steadily worsening. By October 10th the Arab members of OPEC had announced that they would hold their own meeting in Kuwait to decide about using oil as a weapon of war. It was clear that oil and politics were becoming dangerously mixed.

There were no diplomats present, and the critical negotiation was in the hands of the five company men led by Piercy of Exxon and Bénard of Shell. On October 10th, when the Arab-Israel War was at its height, they cabled their boards, who were waiting back at headquarters, that the two sides were still far apart; but there was still a possibility that they could reach an agreement, at somewhere below $5 a barrel. The boards of Exxon and Shell replied firmly; the companies could not go further. The Shell board insisted that the financial effects were so far-reaching that they required consultation with the governments.

After midnight Piercy and Bénard went to see Yamani at his suite at the top of the Intercontinental Hotel. Piercy explained to Yamani that the companies could not negotiate further until they had talked with the consumer governments—which would take as much as two weeks. Yamani was silent for what seemed like ten minutes. Then he ordered a Coke for Piercy, cut a lime with a knife, and slowly squeezed it into the Coke. He was playing for time, hoping that Piercy might keep negotiations going: he was worried that if the question of the price of oil was left unresolved, it would get mixed up with the question of the embargo, which he knew was about to come up in Kuwait. But the company men could not budge. Eventually Yamani said: 'they won't like this.' He rang up Baghdad, talked agitatedly in Arabic, and then told the oilmen: 'they're mad at you.' He rang up the Kuwaiti delegate, who arrived in his pyjamas, for further worried discussion. Then Yamani began looking up airline timetables, still hoping that the oilmen might relent. But they could not. The meeting broke up. When the company men asked what would happen next, they were told: 'Listen to the radio.'

Would history have been different if the companies had been able to settle? The question has since troubled more than one of the delegates; possibly they might have continued to negotiate with OPEC, and never let them take the price into their own hands. As it was, the parting was final: the companies have never since negotiated with OPEC. The buffer was bashed in.

Two days later the companies heard on the radio that the members of OPEC had met again in Kuwait, and decided for themselves to push up the price of oil from $3 to $5 a barrel. The next day the Arab members of OPEC announced the embargo against countries supporting Israel. The coincidence of the two acts provided the catastrophe: the embargo led to the shortage, and thus to the further doubling of prices. And the companies could only look on helplessly, while OPEC apparently usurped their old role.

But were they really helpless? The most critical battle in the story was still to come. For the companies now watched the short-age of oil giving way once again to a surplus, as it had done so many times before. While OPEC still managed to hold out for the high price and even increase it, the fragility of the cartel became more apparent: and the United States government was now hectically trying to organise a consumers' protective organisation, or counter-cartel. There were now, as there had been at the very beginnings of the industry, two rival potential cartels, the con-sumers' and the producers', each keenly watching the other, for the weakest members to give way. But whose side were the com-panies really on? Did they want to break the OPEC cartel, or join it? Had they already effectively joined it, by running the regulated system for the countries, making sure that no one country should flood the market, and bring prices tumbling down? It was a conflict without precedent, but it was also the last of a whole succession of battles for the control of the oil industry over the last century. In this book I try to trace episodes in the long global history that led up to these last sudden tilts of the see-saw, and to show the interplay of relationships between the governments and companies. The story begins with the dramatic birth of the industry in Pennsylvania, the first battles between producers and distributors, and the rapid victory of the Rockefeller monopoly, which grew beyond the reach of governments. It passes on to the heirs and rivals of Rockefeller, and the autocratic men who carved up the oil world between them and worked out their cartel to regulate it. Then it moves on to the seven post-war corporations which shared out the great new reserves in the Middle East, gradually changing the centre of gravity of diplomacy, and balancing countries and rulers against each other. Then it crosses over to show the emergence of the rival cartel, the new club of OPEC, slowly mobilising itself against the companies until the two sides were arrayed for the climax.

In depicting this long global conflict, I concentrate on the decisive battles and turning points—including Iran in 1954, New York and Baghdad in 1960, Libya in 1970, Vienna in 1973—which illuminate brightly the issues and policies. I try to reconstruct the problems and choices, taking the bones of my evidence from documents and testimony, and filling in the flesh and blood with interviews with the characters concerned, to depict the collision course from both sides.

Through the narrative I try to raise the questions and might-have-beens that recur through oil history. How did these great companies, more completely than others, grow up with so little control from their home governments? What was the relationship between this slippery industry and the conventional processes of diplomacy? Could that sudden shift of balance of 1973 have been more gradual and less drastic if the companies had shared or broadened the base of their power? Above all, why and how does oil so persistently fall into the hands of either a monopoly or a cartel—first of Rockefeller, then of the seven sisters, then of OPEC? And in whose hands will it fall, or should it fall, in the future?

NOTES

1. *Forbes Magazine*, April 1, 1973.
2. Reviewing my book *The Sovereign State of ITT* in the *Sunday Telegraph*, July 14, 1973.
3. J. E. Hartshorn: *Oil Companies and Governments*, London 1967, p. 378.

The Rockefeller Inheritance

The American Beauty rose can be produced in all its splendor
only by sacrificing the early buds that grow up around it.
John D. Rockefeller, 1905

Down with all tyrants! God damn Standard Oil!
Phil Hogan in Eugene O'Neill's play
A Moon for the Misbegotten, 1953

Most of the problems of the oil industry were already evident in
the first few years of its existence, when it suddenly grew up a
century ago in Pennsylvania. The alternation of shortage and
glut; the hectic oscillation of prices; the battles between producers and distributors; the interdependence of oil and transport;
and above all the question: who should control it?

To search out its origins I visited, like many pilgrims before me,
the Oil Regions in Pennsylvania where the business first began;
and that journey has ever since stuck in my memory. It is a
beautiful countryside: the trail to the oilfields led north from
Pittsburgh, along the Allegheny River, with the grime of the city
giving way to green hills on either side. Approaching the Regions
there was little sign of any industry at all, past or present; only a
few barns painted with faded advertisements for a chewing
tobacco called Mail Pouch, and the picturesque old Wolf's Head
refinery on the river with a high chimney belching smoke. Further
on, I came to Oil City: a sedate little town looking prim and prosperous with a twin-spired church on the hill, some fine Victorian
buildings and a new Holiday Inn along by Petroleum Street.
There were few indications of its past frenzy: the bookshop had a
shelf of nostalgic reminiscences and books of photographs of the
town in the oil boom, written by local oil buffs; but there was
little to connect the modern town with the old pictures of seedy
characters lurking in the muddy streets outside shanty hotels.
Beyond Oil City, away from the river, the country became

wilder, and the road more winding, until a signpost pointed to Pithole. It was the most legendary of the oil-rush cities; at its peak, in 1865, it had 10,000 inhabitants, eight hotels, two telegraph offices, a theatre, a daily newspaper, and (as one contemporary described it) 'fifty free-and-easies, affording as big a den of vice as the world has ever seen'. At the time, it was the centre of oil communications: the Pithole post office was said to be the third busiest in the States. The first pipeline in history was laid there, to carry oil six miles to the river; and a railway extension was built, linking Pithole to Oil City and Oleopolis.

WHERE THE OIL INDUSTRY BEGAN

But arriving at the spot, there was nothing: only a notice on top of the hill proclaiming the Ghost Town which lived for only five hundred days, and a museum under construction; there were the remains of an old waggon, a few drive-pipes in the grass, and a picnic place for sentimental travellers. No sign of the main street, the theatres or churches: after the town was abandoned they blew down, or caught fire, or were dismantled to build new boom towns.

I drove on, through Oleopolis and Pleasantville, alongside the route of the first pipeline, to the small town of Titusville, where oil was first drilled in 1859, by 'Colonel' Edwin Drake: outside the town the Drake Well Museum was been built as a belated memorial to the poor pioneer: like so many other oil heroes, he died a pauper, while the entrepreneurs got rich in his wake.

Inside, Drake is depicted in effigy, a fine-looking bearded figure with a top hat, and outside is a replica of his rig, a simple wooden contraption with the 'walking beam' moving up and down. Close by is Oil Creek, which used to be the artery of the oil industry, with oilfields and towns along its filthy length. Now it is perfect for a picnic.

Walking through the green fields of the Oil Regions, comparing them to the old photographs, the first oil age of a century ago seemed as remote as the civilisation of Pompeii. Yet its consequences were all around me—cars, electricity, fertilisers, all dependent on oil. The Ghost City was a useful reminder of the most persistent worry about oil; it runs out.

But the early pioneers were confident that there would always be oil somewhere else. To many of the early adventurers, there was something providential in the discovery of the precious fluid, just when the whales which had provided the lamp-oil were beginning to disappear from the oceans. The shack towns might come and go, but the drillers did not doubt that there would be new ones, and as the Pennsylvania pioneers spread out through America the oil business soon acquired a special nomadic character. Only a few years after Drake's drilling, America without oil seemed unimaginable: as one reporter, J. H. A. Bone, described it in a book in 1865:

> From Maine to California it lights our dwellings, lubricates our machinery, and is indispensable in numerous departments of arts, manufactures and domestic life. To be deprived of it now would be setting us back a whole cycle of civilisation. To doubt the increased sphere of its usefulness would be to lack faith in the progress of the world.[1]

On this optimistic expectation, the transformation of the Western world began. As time and again a new shortage gave way to a new glut, the assumption was wonderfully vindicated: so that it became impossible to visualise a world without unlimited oil.

ENTER THE BOOK-KEEPER

The men who flocked to the Oil Regions in the first years, many of them returning veterans from the Civil War, soon discovered another fact about the new industry: that it was subject—far more than coal, or gold—to disastrous over-production, leading to sudden collapses in the price. It was here that the ominous phrase was

first coined: 'the bottom fell out of the market'. The lives of the
drillers turned quickly from excitement to misery, as the prices fell.
The year after the first discovery, the price of oil was $20 a barrel;
at the end of the next year it was 10 cents a barrel, and sometimes a
barrel of oil was literally cheaper than a barrel of water. (The
barrel, equal to 42 U.S. gallons, which has been the chief measure
of oil ever since, had its origin in the wooden barrel from Pennsylvania.) The vivid reports of contemporary travellers, and the
wonderful early photographs of the regions, taken by John Mather
of Titusville, record the desolation and anarchy in the regions:
sleazy men in waistcoats and bowler hats sit beside bits of old
machinery, derricks crowd up against each others' legs, and rivers
of mud, oil and rubbish flow through the towns.

But there was another idealised view of the pioneers' lives, for
they represented a heroic ideal of self-reliance and free-enterprise.
However wretched their condition, they could always—like prospectors everywhere—hope for a magnificent future. The world of
the oil drillers acquired in retrospect a legendary romance whose
splendour was eloquently expressed by one of the children of the
Oil Regions, Ida Tarbell, whose father was driller at Pithole.
Writing forty years later, in the classic book on Standard Oil,
Tarbell looked back on an idyllic world of her childhood:

> Life ran swift and ruddy and joyous in these men. They were
> still young, most of them under forty, and they looked forward
> with all the eagerness of the young who have just learned their
> powers, to years of struggle and development. They would solve
> all these perplexing problems of overproduction, of railroad dis-
> crimination, of speculation. They would meet their own needs.
> They would bring the oil refining to the region where it belonged.
> They would make their towns the most beautiful in the world.
> There was nothing too good for them, nothing they did not hope
> and dare. But suddenly, at the very heyday of this confidence, a
> big hand reached out from nobody knew where, to steal their
> conquest and throttle their future. The suddenness and the black-
> ness of the assault on their business stirred to the bottom their
> manhood and their sense of fair play, and the whole region arose
> in a revolt which is scarcely paralleled in the commercial history
> of the United States.[2]

For Ida Tarbell, and for many of the pioneers, the Big Hand
was represented by one man: Rockefeller. Among Mather's early
photographs, there is one taken in 1864 which has a special fasci-

nation. It shows a cluster of derricks and ramshackle huts in Oil Creek, with a group of men, all in bowlers, standing round a stagnant pool. In the foreground, sitting on a bank, is a young man more formally dressed in puttees and a long coat with a wide-brimmed hat, looking down on the scene. He is said to have been identified as John D. Rockefeller. Certainly the picture represents vividly enough the crucial characteristic of the man: his detachment.

Only six years after the discovery of oil the young Rockefeller —then only twenty-six—had bought control of a refinery business in Cleveland, a hundred miles west of Oil City. Already he understood the critical facts of the new industry, and he had the capacity of cold analysis which was to mark all the great oil tycoons. He had small eyes, high cheeks and a long mask-like face, which expressed his dread of emotionalism. His parents had forged an iron discipline; his mother, a devout Baptist, would tie him to a post and beat him when he was disobedient, while his father, a bogus doctor who sold patent medicines, would trade with his boys and cheat them, to 'make 'em sharp'. 'Doctor' Rockefeller was an extrovert character of doubtful reputation who was once, rightly or wrongly, indicted for rape, a background which doubtless encouraged his son to withdraw into himself and into obsessive hard work. As a young book-keeper Rockefeller had to promise himself not to work after ten in the evening, so fascinated was he by figures. He was adept at mental arithmetic, and was proud to discover that by quick calculation he could 'beat a Jew'. As he surveyed the chaos of the Oil Regions he was appalled by the disorder and instability of the new industry, and he despised the careless camaraderie of the producers; in later life he warned young men 'don't let good fellowship get the least hold of you'. Above all, he was secretive and taciturn, and liked to recite a doggerel quatrain, which could be the motto of the industry he created:

> A wise old owl sat on an oak;
> The more he saw the less he spoke;
> The less he spoke the more he heard;
> Why can't we be like that old bird?[3]

In Cleveland, after beginning as a book-keeper, Rockefeller went into partnership in a refinery with two easy-going Englishmen, the Clark brothers, but soon bought them out. He expanded with great daring, borrowing wherever he could, and bringing in new partners. He realised that the only way to dominate the industry was not by producing oil, but by refining and distributing it, and undercutting his rivals by cheaper transport. With the help of a new partner, Henry Flagler, he persuaded the railroads to give secret rebates to his oil, extending the existing practice of allowing discounts for large quantities of freight. Was it Rockefeller who destroyed the spirit of free enterprise, or was it the railroads? Oil and railways grew up together, and before oil had ever been discovered, the Atlantic and Great Western Railway had been planned, connecting the oil regions to Cleveland and extending a branch line into Titusville.[4]

'We are at last in America' wrote the *Pithole Daily Register* in 1865. It was his proximity to the railroad which gave Rockefeller his great opportunity. The Atlantic and Great Western was to become the chief carrier of oil in the country, and with the railroads came the expansion of the industry to a continental scale.

Rockefeller saw nothing wrong with exploiting the system of railroad rebates to the limit. 'Who can buy beef the cheapest?— the housewife for her family, the steward for a club or hotel, or the commissary for an army?[5] But it was the secrecy of Rockefeller's methods, as much as their ruthlessness, that made him such a special figure of hatred. As his trade and refineries expanded his rivals never knew quite what was hitting them; when he bought out his Cleveland competitors, he kept the fact secret, so that they pretended to be still competing, thus serving as spies among other genuine competitors. By 1870, after only seven years in the business, he was able to establish a joint-stock company called the Standard Oil Company with a capital of a million dollars, of which he himself owned 27 per cent. Already the company held a tenth of the oil industry in America.

Two armies were set for the first of many classic battles for the control of the industry. The producers and drillers in the Regions, on one side, had been decimated by the overproduction and depression, and were determined to organise themselves to control production. On the other, Rockefeller and his fellow-refiners had a firm hold over transport. They first tried to organise the notorious 'South Improvement Society' which gave handsome railroad rebates to those who belonged to it, and penalised those who did not. The producers in turn formed their own committee and counter-attacked, boycotting the cartel, and eventually forced the railroads to give them cheaper rates. For a time it looked as if the producers *could* stand together. They agreed to stop new drilling, to sell oil at a fixed price, and reached an understanding with the refiners. But as the glut of oil continued the independent producers could not resist undercutting, and the market once again collapsed. Rockefeller was convinced that the reform of the industry could only be achieved by his own company: 'we proved', he wrote in his memoirs, 'that the producers' and refiners' associations were ropes of sand.'[6]

By 1875 the refiners had become more effectively organised, with Standard Oil as the leader. The railroads had combined to share the traffic and the refiners had made their own Central Association, with Rockefeller as its president. From then on, it was much easier to persuade other refiners to sell out to Standard, which provided a haven against the competitive storms, and which gave the prospect of steady profits and security. Rockefeller saw his victory as the result not of his own ruthlessness, but of the hopeless weakness of the producers who could not control their own industry. In a sentence that is still relevant to producers today he explained: 'The dear people, if they had produced less oil than they wanted, would have got their full price; no combination in the world could have prevented that, if they had produced less oil than the world required.'[7]

For Ida Tarbell, representing the producers, the overproduction was a natural phenomenon, like a wild harvest: 'It seemed as if Nature, outraged that her generosity should be so manipulated as to benefit a few, had opened her veins to flood the earth with oil.' But for Rockefeller there was nothing natural about it: 'It is to be remembered,' he said in reply to Tarbell, 'that Nature would not have opened her veins if the producers had not compelled her so to do.'[8]

The producers, weakened by their own overproduction, were routed while Rockefeller gobbled his rivals. By 1883 he had formed the Standard Oil Trust on a continental scale. His advantage on the railroads was now less important, for his company had its own network of pipelines, the 'iron arteries' pumping the product through the Eastern United States. The pipelines, which had first been developed by the smaller producers as part of their defence were now the most effective instruments of the monopoly. Through the device of a trust, which held shares in each component company, Rockefeller was able to circumvent the laws which then prohibited a company in one state from owning shares in another; at the same time he could and did pretend that all the companies were independent. He had no personal doubts about the rightness and need for this kind of concentration: 'This movement', he said later, in a famous passage, 'was the origin of the whole system of modern economic administration. It has revolutionised the way of doing business all over the world. The time was ripe for it. It had to come, although all we saw at the moment was to save ourselves from wasteful conditions . . . The day of combination is here to stay. Individualism has gone never to return.'

From his headquarters at 26 Broadway in New York Rockefeller controlled a corporation unique in the world's history. It was almost untouchable by the state governments which seemed small beside it, or by the Federal Government in Washington, whose regulatory powers were still minimal. By bribes and bargains it established 'friends' in each legislature, and teams of lawyers were ready to defend its positions. Its income was greater than that of most states (as the income of modern multinationals is bigger than most nations where they operate). Its profits were big enough to finance its expansion, independent of the bankers whom Rockefeller resented. As the oil industry grew, and moved on from Pennsylvania to Ohio, Kansas and California, so Rockefeller began to buy oilfields as well as refineries, bringing Standard Oil closer to self-sufficiency and to the 'integrated' oil company of today. Its oil was already exported to Europe, to the Middle East and to the Far East, providing kerosene for oil-lamps, and fuel for stoves and for ships. By 1885 seventy percent of the Standard's business was overseas, and it had its own network of agents through the world, and its own espionage service, to forestall the initiatives of rival companies or governments.

It was itself shrouded in the secrecy which was to be characteristic of oil companies: 'if there was ever anything in this country' wrote J. C. Welch in 1883, 'that was bolted and barred, hedged round, covered over, shielded before and behind, in itself and all its approaches, with secrecy, that thing is the Standard Oil Company.[9] However much the company was attacked and abused Rockefeller had insisted that his colleagues should not reply, so that—as he realised years later—'the more we progressed yet kept on gaining success and keeping silent, the more we were abused'. 'I shall never cease to regret,' he said, looking back at the first major attack on his company in 1872, 'that at that time we never called in the reporters.' But the secrecy was an essential part of Rockefeller's expansion, and at the time Rockefeller himself appeared impervious to public attack. His closest colleague, Henry Flagler, once burst out: 'John, you must have a hide like a rhinoceros!' He bequeathed that hide to the industry.

THE COUNTER-ATTACK

Gradually the public and the politicians were catching up, and realising the extent of this and other monopolies. One of Rockefeller's historical achievements was to provoke the anti-trust movement and the machinery of Congressional investigation which has pursued the American oil industry ever since. The tide of public opinion was turning against the great new industrial corporations, and already as early as 1871 Charles Francis Adams was describing how corporations 'have declared war, negotiated peace, reduced courts, legislatures, and sovereign states to an unequalled obedience to their will'. A succession of muckraking writings made Rockefeller into one of the most unpopular men in the States, beginning with a savage article by Henry Demarest Lloyd in *Atlantic Monthly* in 1881, and reaching a climax in 1904 with the *History of Standard Oil* by Ida Tarbell. The muckrakers helped to stir first the states and then the Federal Government. So began the first of the conflicts between the oil industry and government which have recurred in cycles over the twentieth century.

The Republican Senator John Sherman proposed a bill to 'declare unlawful, trusts and combinations in restraint of trade and production' and the Sherman Anti-Trust Act was signed in the summer of 1890 by President Harrison. With deliberate vagueness it prohibited 'every contract, combination ... or conspiracy in restraint of trade'. It was a historic milestone, as the first major counter-attack by the federal government against corporate monopoly, and the beginning of a movement which had no parallel in Europe. But its limitations were evident to many at the time. It was designed to reassure the public more than to take effective action, and it was followed by a long period of Federal inaction. Anti-trust from the beginning had a bark that was worse than its bite.

Many states in the meantime were passing anti-trust laws. The Attorney-General of Ohio brought a bold suit against Standard Oil, which was upheld by the Ohio court. Rockefeller however could now take advantage of the State of New Jersey, which since 1888 had new laws which allowed corporations to hold shares outside the state. The trust was thus now reconstituted round a holding company called Standard Oil (New Jersey) which owned shares of all the other components of the trust.

It was under Theodore Roosevelt that the anti-trust laws became seriously effective, when a special anti-trust section was set up with five lawyers. In 1906 Attorney-General Moody announced that he would bring to trial every possible case against Standard Oil. A massive suit was filed against the company under the Sherman Act, and in May 1907 a Report was published by the Commissioner of Corporations. It remains today a fascinating study of the workings of an industry, tracing the operations of the monopoly throughout the Union.

A special prosecutor, Frank Kellogg, amassed the detailed evidence of Standard's monopoly and exorbitant profits—nearly a billion dollars in a quarter-century—and Rockefeller himself appeared to give testimony. Now an impassive old man of sixty-nine with hollow cheeks and a white wig, he had long since ceased to run the company, but he was still nominally president and responsible for its methods. He described to his own lawyers how he had benignly built up his great empire, but as soon as Kellogg began cross-examining, as one newspaper noted, 'his mind was opaque as an oyster-shell'. The circuit court upheld the government, and Standard appealed to the Supreme Court, which handed down its historic decision in May 1911. Chief Justice White read his 20,000 word opinion, describing how the 'very genius for commercial development and organisation' which had created the Standard Oil trust had pressed it to form a monopoly, and to 'drive others from the field and exclude them from their right to trade'. The Supreme Court decreed that within six months Standard Oil must divest itself of all its subsidiaries.

INEVITABLE MONOPOLY?

The thirty-year wonder of the Rockefeller monopoly has raised questions which have been argued by historians and economists ever since, and which have obvious relevance to the subsequent history of oil and the current crisis. Was it the ruthlessness of the individual that set this huge industry in the pattern of monopoly? Or was it of the nature of oil, unlike other industries, that it could never survive as a free enterprise industry ruled by the normal

laws of supply and demand, and that it was bound to be controlled
and restricted by someone—whether by Rockefeller, or by a cartel
of a few companies, or by OPEC?

The second view has been cogently argued by Paul Frankel, the
London oil consultant and economist, who maintains that the oil
industry is subject to laws quite unlike most others: and his argument has gained wide support. In most industries the problem of
overproduction, leading to falling prices, is solved by the workings
of the market; as prices come down, so some companies go out of
business, production falls, and a balance is reached between supply
and demand. But oil, Frankel maintains, is quite different. In the
first place price variations, except in extreme cases, have very little
effect on demand: people need oil so badly, whether for lighting,
heating or cars, they they will pay far more than it costs to produce. In economists' terms, it is price inelastic. In the second place
lower prices often have little 'effect on limiting production. The
drillers in Pennsylvania still went on selling their oil at a few cents
a barrel, so long as there was oil underneath; and in later years the
big companies with huge concessions still went on producing at
uneconomic prices, to keep their refineries going at almost any
cost.

The industry thus failed to be self-adjusting, and was bound to
be controlled either by a single company or some kind of cartel. In
Rockefeller's case it was controlled by the command of the crucial
bottlenecks—first railroads, then pipelines. In Frankel's words:

> What began in Pennsylvania of the seventies developed ac-
> cording to a pattern which we can detect in the oil industry all over
> the world down to this very day: the ascent of one concern or of a
> group of concerns which, by centralisation of control and by
> dispersion of interest, attains in due course a paramount position.[10]

Frankel wrote that in 1945, but he still confidently sticks to his
position in 1975: 'I am a cartelist,' he says.

Many economists have since agreed with Frankel, but others
have strongly dissented, and the question has become a highly-
charged one among the oil experts. Professor Maurice Adelman,
the fierce critic of the oil companies from the Massachusetts
Institute of Technology, insists that in competitive conditions
over the long term, the oil industry behaves like any other; the
greater the output of an oilfield, the higher the cost of additional
output; so that the 'crude-oil industry, contrary to common

belief, is inherently self-adjusting'.[11] Professor Edith Penrose, another formidable critic of the companies from the School of Oriental and African Studies in London, largely agrees with Adelman.[12] The notion of a 'natural monopoly', Adelman suggests, was 'a scholarly gloss on revealed truth. When half a dozen companies made up the international industry, they *knew* that they should be allowed, statesmanlike, to divide markets and set prices'.[13]

If Adelman's view is taken, Rockefeller and his legacy cannot be regarded as an inevitable historical development. The oil business *could* have been divided between genuinely competing companies. But even if control *was* inevitable, a further question remains open; who should be the instrument of that control— producers or consumers, companies or governments? The epic battle between the Pennsylvania producers and Rockefeller, and between Congress and Standard Oil, would be repeated in many different forms over the next decades. The fascination of oil history lies in the changing form of the battle to control the supply.

THE LEGACY

The Rockefeller inheritance left a long trail. In the first place the transfer of wealth achieved by the oil monopoly—as spectacular in its time as the shift of funds to the OPEC cartel today—served to establish a system of philanthropy and foundations which provided a kind of secondary government, influencing the direction of research and education. In its separateness from the government, its immunity from taxation and its concern for quality, the Rockefeller fortune had some resemblance to the estates and patronage of the royal families of Europe. According to his grandson Nelson, John D. Rockefeller gave away a total of $550 million to various foundations, and gave to his only son $465 million: while paying only $67 million in taxes (income tax was only invented in 1913). His son John D. II gave away $552 million, paid $317 million in taxes, and gave to his family, including his five sons, a total of $240 million. Among those sons, Nelson claims to have given away or pledged $53 million and to have paid $69 million in taxes; while the value of his own and his trusts' assets in 1974 was $218 million.[14] In spite of all the benefactions, the total value of the assets of all living descendants of John D. Rockefeller was estimated in 1974 at two billion dollars—which remains, as far as can be ascertained, the biggest family fortune in the world.

The hateful image of John D. Rockefeller I was gradually transformed with the help of the first public relations man, Ivy Lee, who cultivated the notion of a benign old man giving away dimes and preoccupied with charity, and supervised his appearance to the press.[15] To the contemporary reader there is an impossible contradiction between the two faces of Rockefeller—the unscrupulous monopolist on the one side, the pious philanthropist on the other. But there is no indication that Rockefeller saw any disconnection. In each case, he was serving as God's instrument and imposing order upon chaos. The new race of master-industrialists, of whom Rockefeller was the first, remained confident that they were in the process of building modern America, forging the links and levers across the continent long before the state and federal governments had comprehended its scope. They regarded the critics and trust-busters as nostalgic and economically ignorant—rather as sixty years later modern multinational tycoons would complain about the objections of nation-states. 'The struggle against Rockefeller's movement for industrial consolidation,' writes Nevins, 'was not a struggle against criminality; it was largely a struggle against destiny.'[16] And Rockefeller's grandson Nelson echoes Nevins' view, that 'the size of his fortune was an historical accident'.[17]

Upon the oil industry, Rockefeller left an indelible mark. Even after the original monopoly was broken up and the family became aloof from the business, the methods and attitudes of the old man inspired his successors. It was not only that he established a system of organisation and control over the anarchic industry and a common understanding which was bequeathed to each of the dismembered components. It was also that behind the fortress of 26 Broadway he established the separateness of the industry, defying governments and societies, vulnerable only to the most extreme legal sanctions. His lack of scruple and his mendacity provoked a continuing distrust of the oil industry, by the public and politicians, which was never to be allayed. Part of it was well-justified, but part of it arose from the need for a scapegoat to blame for large-scale organisation. The first epic battle between the oil-drillers and Standard Oil for the control of the industry, as so eloquently described by Tarbell, became a kind of morality play for any individual who felt threatened by the new industrial order; a kind of commercial equivalent to the Western, as the last glimpse of a heroic age.

The anti-trust movement, the sheriff in the Western, was a force not to be underestimated, for it was supported not only by the theory of free enterprise but also by a faith in the power of the individual against organisations. The trust-busters saw themselves as defending the very core of democracy. In the next six decades there were to be successive duels between the oil companies and the Justice Department, and most often the oil companies would win. But the duels were played out before a public opinion which could never be wholly reassured about the giant combinations. Periodically, in sudden gusts of anger which recur through this story, they would turn on the oil companies as the symbols of everything that was sinister and secretive in modern industrial society.

THE STANDARD SISTERS

The dissolution of Standard Oil was in many ways drastic, returning each component to the state where it operated. But the thirty-eight companies were still owned by the same group of men and led by Rockefeller himself with a quarter of the shares. The companies preferred not to trespass on their rivals. The price of oil actually went up. The dividends increased and the shares of most of the companies rocketed—partly because shareholders now had more information, but partly because of a massive new boom in the oil trade. There were some critics—including the young Walter Lippmann—who believed that it would have been more in the public interest to have left the monopoly and controlled it firmly from Washington. The break-up looked like a radical reaffirmation of free enterprise, but it funked the problem of the government's control.

Of the thirty-eight progeny some were soon to become bigger, in terms of profit, than the original parent, as the oil industry expanded to provide fuel for industry, and gasoline for cars, planes or tanks. Three of the offspring were soon to be among the seven sisters who make up the main story of this book.

STANDARD OIL OF NEW JERSEY (EXXON)

By far the biggest of them was the rump of the original New Jersey company which had acted as the holding company for the rest. This company, now known as Exxon or Esso, has for the

following sixty years maintained its position as the biggest oil
company in the world, setting a style for the rest. After the dissolution of 1911 it seemed in many ways to have changed very little
in its fortress on Broadway. It still bought most of its oil from the
other Standard companies, and being much richer in assets it still
served as the banker to many of them, maintaining their loans for
exploration and development. It still had an unparalleled network
of outlets in America and the world. And it inherited many of the
directors who had helped to build up the Rockefeller system. At
the head of them was John D. Archbold, who had virtually been
running Standard Oil for fifteen years before the break-up, after
Rockefeller had disengaged himself from the business. For two
decades this effervescent and relentless little man was virtually king
of the oil industry.

Archbold was an Irishman, the son of a Methodist minister, who
had begun selling oil in Pennsylvania. Like many of Rockefeller's
later colleagues he was at first a bitter rival, campaigning vocifer-
ously at Oil City against Standard, but he soon secretly sold out
and became Rockefeller's favourite and a director of Standard at
only twenty-seven. He was a jovial, popular man who was said to
have 'laughed his way to the top'; he was a heavy drinker, until
Rockefeller insisted that he should sign the pledge. But behind
his jocularity he was totally preoccupied with profits and oil: 'he
would make up his mind with one flash of his dark snapping eyes,
and then was smiling again'.[18] He was more purely an oilman than
Rockefeller, untroubled by doubts about other interests, and he
was more contemptuous of governments: asked by the New York
State committee about his role in 1879, he had replied, 'I am a
clamourer for dividends. That is the only function I have in
connection with the Standard Oil Company' (a remark which
Rockefeller in his memoirs quotes as evidence of Archbold's 'well-
developed sense of humour').[19] He clamoured loud, and the more
exorbitant profits of Standard Oil dated from his rule. He also
established a network of bribes (eventually uncovered by the
Hearst Press) to Senators and Congressmen to safeguard the
monopoly.

Under Archbold Exxon emerged as a separate entity. The other
directors included many of Rockefeller's early associates, now all
multimillionaires, including Charles Pratt (whose old mansion
now houses the Council of Foreign Relations in New York) and
Edward Harkness (who bequeathed, among other gifts, the Com-

monwealth Fund, which has financed generations of British students visiting America). The company lost little of its arrogance, and the directors, lunching together every day on Broadway, remained cut off from government or public opinion. Their contact with the press was a poet called Joseph I. C. Clarke, whose main job was to avoid giving information. The company remained the object of intense distrust as (in the words of the Congressional Record in 1914) 'the invisible government'. When the Federal Trade Commission investigated in 1915 the effects of the dissolution, it concluded that the inter-ownership of shares still effectively curbed competition and that the Standard companies maintained a common 'understanding'.

The dissolution really came at an opportune time, for the old Rockefeller directors were mostly over sixty and the very success of the monopoly had made them complacent. The split-up soon compelled Exxon to become more aggressive once again, for it was cut off from the crude oil which had been provided by other members. It had money, and markets, but very little oil: it was more like a bank or a trading firm than an oil company. In its need to find its own oil Exxon was forced to look outside the United States. So paradoxically it was the break-up which was to turn Exxon into a major force on the world political scene.

MOBIL

Another offspring of Rockefeller's empire was the Standard Oil Company of New York, known from its telegraphic address as Socony. Like Exxon it found itself with huge markets: it was selling oil in England as early as 1882 and it had extended itself through the Far East, where it made oil and lamps for China. But it had no crude oil of its own. Like Exxon it was based on New York, with a more international and sophisticated approach than the others. Its chairman was another of Rockefeller's associates, Henry Clay Folger, a much more literate oilman than most, who spent part of his fortune buying thirty-five Shakespeare first folios, which he bequeathed with the Folger library to the nation. Like Exxon, Socony was soon forced to look for its own supplies of crude oil to become an 'integrated company'. By 1925 it had bought the Texas producing company called Magnolia, which had earlier had a secret connection with Standard Oil. Eventually in 1931, in the wake of the great depression, it was allowed to merge with another oil company—specialising in lubricants called

Vacuum, which had once been secretly allied with Rockefeller.[20] The merger of Socony-Vacuum thus represented a union of two of the old Standard components; the name was later changed to Socony-Mobil, and eventually just to Mobil. With a chequered history and tangled roots, Mobil was the smallest of the seven sisters, and was also consistently hungry for crude oil—a combination which was to give it a special combativeness.

SOCAL

In California, in the meantime, the battle between the pioneers and Rockefeller had repeated itself. The Western oil industry had been set off two years after Drake's first drilling when in 1861 the first oil drillers travelled across the continent, attracted by stories of oil seepages in the empty new state (California had only been bought from Mexico thirteen years before). One of them, Demetrius Schofield was a remarkable example of the nomadic nature of the new industry; he had operated wells in Pennsylvania, travelled to Japan to investigate the kerosene market, worked as an oil broker in New York, and later speculated disastrously on the San Francisco stock exchange. Schofield now formed his own company, and dreamt of supplying oil to the Far East, but the finds were disappointing and the oil was smoky and smelly. In the meantime Rockefeller had already set up a branch office in San Francisco and began sending oil from Pennsylvania by clipper round Cape Horn.

The California producers fought back gallantly against Rockefeller's price wars, but he steadily strengthened his hold. His real opportunity, as in Pennsylvania, came with the finding of a huge new oilfield—this time underneath the city of Los Angeles (a field which still provides many residents with handsome royalties). The overproduction led to the usual collapse of prices, worsened by more cheap oil coming in from Peru. Rockefeller took advantage of the chaos, and in 1895 bought up Schofield's company for less than a million dollars, thus acquiring valuable oil reserves as well as control over the market. Standard Oil was soon able to achieve the ambition of the early drillers—exporting oil across the Pacific to China. California was now part of 'the creeping, crawling black monster'.[21]

But Standard Oil of California or Socal was only part of the Rockefeller empire for eleven years before the dissolution, when Demetrius Schofield re-emerged as an independent boss. The company quickly took advantage of its new-found freedom, expanding its production until in 1919 it provided 26 percent of the total U.S. production—more than any other company. Ever since, Socal has been basically a producing company, in the opposite situation to Mobil—with plenty of oil, but not enough markets. With the railroad, it was now the dominant financial power on the West Coast, fortified behind a ponderous building in the heart of San Francisco. Its attitude to the Federal Government three thousand miles away was disdainful.

Socal had an early trial of strength with Washington on the critical question of the Federal lands in California which had been regarded by oil drillers as open territory. In 1909 President Taft had insisted that three million acres of these lands should be withdrawn from sale, to safeguard the government's oil supplies. Socal fought back, with the help of their influential attorney Oscar Sutro, of the law firm of Pillsbury, Madison and Sutro (which has ever since had a major influence over the company). Sutro cabled the President, whom he knew, and there was lobbying and counter-lobbying in Washington. After the outbreak of the European war the Secretary of the Navy, Josephus Daniels, became the most implacable enemy of the oil companies, insisting on control of special 'naval reserves'. Socal, on their side, had powerful support, including Senator Jim Phelan of California. For ten years the argument continued, until at last in 1919 a compromise was agreed, which allowed relief to oil-drillers already working on government land—a partial victory for Socal. But the question of the control of naval reserves continued to be a bitter bone of contention.

These three daughters of Rockefeller, Exxon, Mobil and Socal, were all for years afterwards called the Standard Oil Group and accused by their many critics of acting in unison. The suspicion was hardly surprising; they all sold their oil at the same price, under the same Standard name, their directors were old Standard Oil men, and their principal shareholder was still John D. Rockefeller. In later years, as we will observe, these three sisters were to come together in their many operations abroad—linking hands, now in one place, now in another, as in an intricate square dance. The indignant denials of collusion were not very credible after the record of mendacity and secrecy of their ugly parent.

But was it a conspiracy, or a common understanding of the harsh facts of world oil? The story which follows poses this recurring

question. What was clear from their origins was that they had
grown up outside any traditional field, whether of government,
banking or earlier industry. They were deeply distrusted, with
some reason, and the men who ran them were cut off behind high
walls from the rest of society.

GULF

The Rockefeller monopoly had already been partially breached
before the dissolution, and the most complete breakthrough was
in Texas. The state was among the first to pass anti-trust laws, and
when they discovered that Standard Oil had a secret subsidiary
selling oil, they threw it out. Standard Oil thus missed out on the
next great discovery which was the birthplace of two of the seven
sisters.

It happened in 1901 at Spindletop, a bleak ten-foot mound near
the coast of Texas, which became one of the most famous names
in oil history. It was given its name when a remarkable engineer
called Anthony Lucas, a former officer in the Austrian Navy,
came to the mound and set up a rotary oil-rig; 'it was the first
great well drilled by an engineer, and it was a well that only an
engineer could have drilled.'[22] To finance his venture, Lucas went
to two prominent oil prospectors in Pittsburgh, Guffey and Galey
who raised a loan of $300,000 from the banking firm of T. Mellon
and Sons.

The Mellon bank, which thus appeared on the oil scene, was a
formidable rival to Rockefeller. Already by the 1860s, before
Rockefeller had established Standard Oil, Thomas Mellon was a
financial power in Pittsburgh, with interests in coal and gasworks;
and his two sons, Andrew and Richard Mellon, became important
financiers. But with their coal interests they were slow to realise
the importance of the oil fifty miles to the north of them and it was
not until 1889, when Rockefeller had already established his
monopoly, that they first tried to challenge him. They bought an
oilfield at Economy, just outside Pittsburgh, which was run by
their adventurous nephew William Larimer Mellon. They expanded quickly selling oil to France and spending a million dollars
on a pipeline to the Atlantic. But eventually in 1895 they were
forced to sell out to Rockefeller. Now at Spindletop only six years
later, they had a new chance and William Mellon was determined
to try his luck again. The risk quickly paid off; on January 10,
1901, Lucas, having drilled to a thousand feet, made a last twist

on his drill and released the biggest gusher then recorded—a black jet of oil which shot up twice as high as the derrick.

With Spindletop the Mellons entered a territory which had not been corralled by Rockefeller, and with apparently bigger potential. Having had their Texas company rejected, Standard Oil was in no mood to move back into the state. The new field was sensational, and four months after the strike Colonel Guffey, who had first backed Lucas, formed a new company with the help of the Mellons. Guffey himself became the president, with nearly half the shares, while the Mellons only had 13 percent, with the rest divided between rich Pittsburgh investors. The directors, most of whom had never seen Texas, wanted to assuage local opinion; since the word Texas was already preempted, they called the company Gulf after the nearby Gulf of Mexico.

James Guffey was a theatrical oilman, who wore a big black hat, stovepipe trousers and loud waistcoats, and he soon became a prominent Texas character. His position seemed impregnable: he controlled the biggest oilfield in the world, and he soon built a pipeline to the coast at Port Arthur, where a refinery grew up above the cow pastures. Soon Gulf ships were carrying the oil to the world and a contract with Shell, the new British company, guaranteed steady markets abroad. But Guffey was a disorganised businessman, and he soon needed further infusions of money. There followed a typical battle between the pioneer and the financier, between Guffey and the Mellons.

The crunch came only twenty months after the first gush of oil and optimism when Spindletop's fountain of oil dwindled to a trickle. It was a disastrous collapse, and Spindletop, like Pithole before it, left not a rack behind. Today the spot is marked—on dubious evidence—by a pink marble obelisk and a replica of a drill, with honeysuckle and juniper growing round. The cows graze round a few old shacks, and the main street of the nearby town of Beaumont, once the centre of the oilmen's night life, is a row of derelict gingerbread arcades and shuttered shops. The money and the vigour have all gone to Houston a hundred miles away.

William Mellon was appalled by Guffey's disorganisation: 'it is very hard for me to be patient with incompetents'.[23] He reported that Gulf would need fifteen million dollars to be profitable. The uncles, unwilling to fork out, once again looked to Standard Oil and offered Gulf to Archbold in New York. But Standard were still smarting under the Texan anti-trust laws, and Mellon was told: 'After the way Mr. Rockefeller has been treated by the state of Texas, he'll never put another dime in Texas.'

The Mellons were thus forced back on their own, and William took over more control as executive vice-president. He believed that Gulf could only compete with Standard if it took advantage of being outside the Rockefeller system of pipelines and refineries; 'Production, I saw, *had* to be the foundation of such business.'[24] (And this has been true of the company ever since.) The exploration gradually produced results and by 1906 the Mellons had another windfall, when huge new reserves of oil were found in Tulsa, in Indian Territory, later called Oklahoma. They needed extra capital to develop it, which gave them the chance to form a new company with the Mellons firmly in command: Guffey was ousted, and bitterly complained: 'I was throwed out.'

Gulf soon became an increasingly self-sufficient company, producing oil in the South West and selling it in the East. Encouraged by the expansion of cars and the demand for gasoline it opened in Pittsburgh the first drive-in filling-station in 1913. Gulf was by now a formidable competitor to the Standard companies, with much greater production than Exxon and with its own fleet, markets and rich banker-owners. But the Mellons knew better, after their earlier experience, than to challenge Standard on their home ground.

TEXACO

Among the hundreds of speculators who rushed down to Spindletop in 1901 there was a tall leathery Pennsylvanian with a rough style and bushy eyebrows known as Buckskin Joe, whose grandfather had escaped the Irish potato famine. Joseph Cullinan had started in oil twenty years before working for Standard Oil; then he moved to Texas, with some secret backing from Standard. At Spindletop he set up his own Texas Fuel Company with a capital of only $50,000. He raised half the money from a syndicate led by James Hogg, the former Governor of Texas (whose daughter became the grand old lady of Houston) and the other half from a shrewd German-born merchant called Arnold Schlaet, who had come from New York to visit Spindletop on behalf of his employers, the Lapham brothers, the controllers of the Leather Trust. The partnership between Schlaet in New York, who knew about finance and marketing, and Cullinan in Texas, who knew about production, was the basis of the new company. They soon formed a larger corporation with the high-sounding name of the Texas Company, later shortened to Texaco. Cullinan, the President, set up offices in a little wooden building with three rooms and a portico in Beaumont.

The Texas Company made good profits by buying cheap Spindletop oil and selling it to sugar planters along the Mississippi, and to Standard in the East. They survived the exhaustion of Spindletop when they found more oil twenty miles away at Sour Lake and by 1904 they were producing nearly 5 percent of all American oil. From New York, Schlaet began to set up a nation-wide sales organisation, to make Texaco less vulnerable to Standard's price-cutting, and by 1908 they even had a tanker called *Texas* sailing across the Atlantic. By 1913 they had built a thirteen-storey headquarters in Houston, already the oil capital of Texas, which still stands, its columns decorated with the Texan star.

Cullinan was proud of the Texan character of the company, and insisted that his staff must 'appreciate a Texas way of life'. He revelled in his personal power and patronage, entertaining grandly, battling against labour unions, and financing Irish causes. Oliver St John Gogarty described him discussing the revival of the *Irish Statesman* in Dublin: 'Mr Cullinan is one of those Americans who are deprived by Nature of revealing their greatness in words. Had I not seen part of his fleet of bronze oil-tankers each as large as the liner to Holyrood, I should never have realised that I was in the presence of a man who could have bought an Irish county.'[25] He was a powerful leader. But the rift between the two ends of the company, in Houston and New York, became wider, and there followed the customary battle between producers and money-men. Schlaet complained to his old boss, Lewis Lapham, that Cullinan regarded New York as the tail of the dog, and that he showed a 'very dictatorial tone'. Cullinan resented Schlaet's conservatism, and his voting against everything Texas proposed; but the extra capital the company needed was mainly supplied from the East. The exchanges became more acrimonious, and in 1913 a whole contingent of stockholders came down from New York to Houston in a special railroad car. Schlaet's men defeated Cullinan who resigned. Schlaet was not considered as successor, whether be-

cause of his 'austere nature'[26] or because there was some doubt
about his Prussian loyalties. Instead the new president was a more
polished Easterner, a graduate from MIT, Elgood Lufkin, whose
father had been one of Standard Oil's agents in the Far East.
Cullinan moved on to run another oil company and to become a
Houston eccentric, flying a black skull-and-crossbones flag from
the top of the Petroleum Building, 'as a warning to privilege and
oppression'.[27] His maverick style left its imprint on Texaco and
its later autocratic commanders; it was ever after the most stubborn
and centralised, always the loner, among the seven.

OILDOM

The five American companies (with the partial exception of
Mobil) had each by 1914 acquired an identity and character that
were to survive over the next six decades. They had grown up
with bewildering speed, first in Pennsylvania, then in Texas, to
establish their separate powers and bureaucracies that seemed to
rival the government's. The Kingdom of Oil, or 'Oildom' as it
came to be called, had become a very separate part of America.
The geographical origins and expertise had encouraged a special
breed, cut off from the centres of politics and diplomacy; the
shanty-towns of Pennsylvania and the swamps of Texas sent forth
rugged, confident men who knew they had transformed the world
around them. Sons followed fathers into the business and aris-
tocracies of oil soon grew up—not only the Rockefellers and
Mellons, but other families, closer to the oilfields, who will recur
through this book; Moffets, Teagles, Lufkins, Warners, Barrows—
an oily saga passing down the generations. The family traditions
widen the gap from the public.

It was a very American industry. Though the companies sold
much of their oil abroad, and built up networks of distributors and
agents, it was the United States that was the source of nearly all
their production. It was not till the First World War that the companies were beginning seriously to look outside America for supplies. But in the meantime there were growing up in Europe two
oil companies who had cornered supplies from the other side of
the world, who were coming into growing conflict with the
Americans. The two sides were arrayed for a much wider conflict,
with deals and double deals across both hemispheres; and the
tactics of Rockefeller were extended and adapted to the global
scale.

NOTES

1. J. H. A. Bone: *Petroleum and Petroleum Wells*, Philadelphia 1865.

2. Ida M. Tarbell: *The History of the Standard Oil Company*, New York, 1904, Vol. I, pp. 36–37.

3. Allan Nevins: *John D. Rockefeller*, New York, 1940, Vol. I, p. 235, and elsewhere for other details of Rockefeller's early character.

4. The promoter of this railroad was a remarkable English entrepreneur, Sir Morton Peto, who has never been given proper credit. He was an engineer whose company built Nelson's column, and then turned to building railways, first in Britain, then across the world. He wrote a book about *Resources and Prospects in America*, which predicted (p. 203) the future interdependence of oil and railways, and the spectacular future of oil. His own future proved less happy; he built himself a grandiose country house at Somerleyton in Norfolk, but his firm went bankrupt.

5. Nevins: Vol. I, p. 304,

6. John D. Rockefeller: *Random Reminiscences of Men and Events*, London, 1909, p. 5.

7. Tarbell: Vol. II, p. 157. See also P. H. Frankel: *Essentials of Petroleum*, London, 1946, p. 71.

8. Nevins: Vol. I, p. 429.

9. Nevins: Vol. II, p. 58.

10. Frankel: p. 77.

11. *Natural Resources and International Development*, Baltimore, 1964, p. 32.

12. Edith T. Penrose: *The International Petroleum Industry*, London, 1968, p. 168.

13. M. A. Adelman: *The World Petroleum Market*, Baltimore, 1972, p. 21.

14. Rockefeller Nomination Hearings (Senate) 1974: p. 44ff.

15. Lee went to great pains to commission the official biography of Rockefeller; he approached famous biographers, including Winston Churchill, who agreed to do the job for $50,000: an offer which Rockefeller turned down. Eventually the biography was written by Allan Nevins, who explained in his preface that he received no subsidy from Rockefeller.

16. Nevins: Vol. II, p. 709.

17. Nomination Hearings: p. 43.

18. Austin L. Moore: *John D. Archbold* (Macmillan) 1930.

19. Rockefeller: *Random Reminiscences*, p. 5.

20. See *Mobil World*: Vol. 32, No. 11, November 1966.

21. Gerald T. White: *Formative Years in the Far West*, New York, 1962, p. 379.

22. Samuel W. Tait: *The Wildcatters*, Princeton, 1946, p. 116.

seemed more extensive than Pennsylvania, and the two Nobel brothers from Sweden—the sons of the inventor of dynamite—acquired a concession. Needing more capital they brought in the French Rothschilds Bank, and the financiers between them began selling Russian oil which soon impinged on the Rockefeller monopoly. In Europe the Nobels and the Rothschilds reached a temporary understanding with Standard and carved up the market between them. But in Asia Rockefeller was determined to maintain his monopoly and the task of challenging him was almost as formidable as in the United States, for Standard Oil, fortified by its huge American profits could afford to undersell in any single market until it forced its competitors out of business.

Marcus Samuel became involved in a syndicate to sell Russian oil, but he soon realised that the only way to withstand a price-war from Standard was to compete in every market at once. Like Rockefeller he could then subsidise low prices in one place with high prices elsewhere. He determined to beat Rockefeller at his own game. Secretly, with help from his trading colleagues, he prepared for his master-stroke. He had storage tanks built in the key distribution points in the Far East and at the same time he ordered a fleet of tankers of a new design which, it so happened, fitted the stringent requirements of the British directors of the Suez canal. As stories emerged in London of this impending new traffic a violent campaign was mounted, apparently inspired by Standard Oil, to stop the tankers passing through the canal. Lawyers lobbied the Foreign Secretary, Lord Salisbury, members of parliament made concerned speeches, and *The Economist* made a dark reference to opponents who objected to the 'Hebrew inspiration' of the scheme.[2] But Samuel was already a man of some standing in the City of London, an Alderman who could count on some government support. The ships were built, the *Murex* sailed through followed by the *Conch*, the *Clam* and many others. By the end of 1893 they were carrying regular cargoes of Russian oil to the Eastern storage tanks, presenting Standard with a simultaneous challenge which, despite their secret agents, took them almost completely by surprise.

There followed a succession of counter-attacks by Standard Oil against the upstart Shell as it tried to reassert its monopoly and its 'perfect order' on the global scene. It dropped the world price of oil still further, thus driving hundreds of small producers and traders out of business. But Samuel and his syndicate with their

growing fleet of ships, their access to Russian oil, and their scattered
trading-stations were still able to survive. Samuel rejected an offer
from Standard to buy him out and in 1897 he formed the Shell
Transport and Trading Company, of which he himself owned a
third, with his family in effective control. The London headquarters of 'Shell' was still tiny with a few clerks and Samuel
relations supervising the trade. It could not compare with
Rockefeller's masterful organisation now led by Archbold and the
'gentlemen of Broadway', with their network of spies and agents,
and their huge internal market. Samuel knew little about exploration or oil reserves and he never had a dominant position in his
home markets in Europe, to which Standard could ship oil cheaply
across the Atlantic. But Samuel was a shrewd trader, with the
experience of London behind him; and he was fortified by growing
profits from the East, especially Japan.

Another smaller oil company also managed to withstand the
price-wars launched from Broadway. This was a Dutch concern
operating in the East Indies which had been given a royal charter
in 1890 and thereafter called itself Royal Dutch. Its origins were
more exotic and adventurous than Shell's, belonging to the world
of Conrad rather than Trollope. It was first built up by a Dutch
trader, Jean Kessler, who went out to Sumatra to supervise a team
of American drillers and found oil. Kessler built up a selling
organisation, with the help of two daring young Dutchmen. One
was Hugo Loudon, a merchant with a long family tradition in the
East Indies; the other was a brilliant young book-keeper called
Henri Deterding, the son of a sea-captain, who had begun in a
bank in Singapore and acquired a complete mastery of oil problems
and a passion for the 'simplification' of figures. He was a short,
restless operator with a magnetic effect on those he worked with,
and at the age of twenty-nine he was put in charge of selling Royal
Dutch oil throughout the Far East.

Royal Dutch soon came up against both Shell and Standard Oil
and the three of them fought intricate battles over sixteen years,
collaborating with other interests under different names, each
plotting with another to destroy the third.

Shell with its fleet, its markets and its capital appeared much
stronger than Royal Dutch. When the news of the Spindletop
gusher came to London in 1901 Samuel pulled off another
apparent *coup*. He quickly negotiated with Colonel Guffey the
founder of Gulf. With singular casualness he signed a contract to

Oil and Empire

Look out upon the wide expanse of the oil regions of the world!
Two gigantic corporations—one in either hemisphere—stand out
predominantly . . .

Winston Churchill, June 1914

The British, from the beginning, were acutely conscious of their
vulnerability in the new age of oil. They had no oil in their own
country (excepting the tiny but resourceful Scottish shale oil
industry) and were dependent on finding it in remote parts of the
world and transporting it long distances home. Transport was thus
even more crucial than in America. It was appropriate that while
Americans always measured oil by the barrel—the measure of the
Pennsylvania producers—the British persisted in measuring it in
tons—the measure of shippers. For the British oil was a long-
distance industry which acquired from the beginning an association with national survival and diplomacy, and oil soon seemed
part of the empire itself.

There was a greater sense of respectability attached to the two
British-born sisters than to the American. The British public had
no searing domestic experiences of the exploitations of oil companies where ruthlessness, like that of slave companies in previous
centuries, was at the other end of the world. There were no anti-
trust cases, no total confrontations between companies and governments, and the Europeans were more tolerant of cartels. The
British oilmen gave the impression that they were less concerned
with profits than with serving the empire and they were rewarded
accordingly. As one American oilman, Robert Wilson of Amoco,
was to complain many years later, 'In England they knight their
leading businessmen, over here they indict them.'[1] The Europeans,
like the Americans, soon had their oil dynasties: the Strathalmonds of BP, the Loudons and Barrans of Shell. But they were
an aristocracy less separate from the traditional powers of government and diplomacy.

Yet the British companies, for all their dignity and imperialism,

were in the end to face the same basic questions as the Americans. As they competed more closely, like all competitors, they resembled them more closely, while governments on each side of the Atlantic looked across to the other for solutions to their problems. On both sides, governments could not decide whether they dared leave such a precious commodity in the hands of businessmen. Should they support them, investigate them, or control them? The arguments within the British government sixty years ago, and the policies of Winston Churchill on control of oil, still have, as we shall see, a relevance today for Britain and for the rest of the world.

SHELL

Rockefeller's oil monopoly, dented by Gulf and Texaco in America, had already been partially broken outside America as early as 1892, nineteen years before the break-up of Standard Oil. The attack began when a new kind of tanker designed to be safe enough for passage through the canal sailed through Suez with a cargo of oil from the Russian oilfield at Batum bound for Singapore and Bangkok. The ship was called the *Murex*, named like all its successors after a sea-shell. The promoter of this brave enterprise was Marcus Samuel, the founder of Shell. He was an entrepreneur who personified the ambivalence of the British oil industry, never sure whether he was concerned simply with profit or with public service.

Samuel came from a much more international background than Rockefeller. Unlike any of the other founders of the oil industry he was a Jew, with a Jewish sensitivity to the world around him and a cosmopolitan home. While Rockefeller was establishing Standard Oil in Cleveland young Marcus Samuel, a shy introverted boy, was being brought up in an austere, close family in the East End of London. He soon inherited a modest fortune from his father, who had traded in the Far East specialising in the fashionable boxes made of shells, which he imported to Britain. Marcus with his brother Sam and some cousins developed and extended the Far East business and very early, like Rockefeller, perceived the crucial importance of economies in transport, looking always for the closest source of supply to meet local demands. He established a strong base in Japan, from which he shipped coal. From coal he began to turn his attention to the new fuel, oil.

He found it in Russia. The Tsar after 1873 had allowed foreign interests to prospect for oil in the Caucasus, a source which already

take 100,000 tons of oil a year from Gulf at a fixed price, for twenty-one years: amounting to half Gulf's total production—the first of many such sisterly agreements. Soon the Shell tankers were bringing in oil from Texas to Europe and Samuel was able to establish a company in Germany, a Standard Oil stronghold. It was a powerful alliance, between Texan producers and a company of world traders.

Standard were now still more determined to buy out this dangerous intruder. Archbold invited Samuel to New York, wooed him and flattered him and offered him $40 million for the business, promising a joint subsidiary with Samuel as chairman. But Samuel preferred to pursue the other alternative of joining with Royal Dutch, where he thought he could be the master. He therefore agreed with Deterding to share Shell's marketing system with Royal Dutch in the Far East in a joint company called the Asiatic, and rashly allowed Deterding to be the head.

The tide, however, soon turned swiftly against Samuel. First Spindletop dried up and Gulf could not fulfil their contract. Andrew Mellon, the chairman, came to England: Samuel could have sued him, but instead agreed to cancel the contract. The Shell tankers were converted into cattle-boats and Samuel was faced with a critical shortage of oil. Then Archbold, after his failure to buy Shell, unleashed a succession of new price-wars broken by short-lived truces. Samuel also faced increasingly formidable rivalry from Deterding, who was rapidly consolidating his position in the Far East with the help of new oil discoveries.

Samuel had by now been increasingly distracted from his business by civic pomp and the life of a country gentleman. He was the epitome of the Edwardian parvenu, a symptom of Britain's industrial decline. He was stout and portentous, a figure out of Belloc. He had bought a large house in Portland Place and a great pile in Kent called the Mote, and he was now Sir Marcus, the first oil knight. In the year when he was challenged by both Deterding and Archbold he was elected Lord Mayor of London—only the third Jew to be so—and his Lord Mayor's show and his banquet were of unparalleled splendour. For Sir Marcus it was the peak of his career, the final mark of acceptance. But his rival Deterding, who was his guest at the banquet was satiated by the pageantry: it reminded him of a Dutch circus.[3] Sir Marcus went back to business after his Lord Mayor's year, to what was described by the press as an empty Shell.

Deterding, thirteen years younger, had the ruthlessness and concentration that Samuel now lacked. Though Royal Dutch was much smaller it controlled valuable reserves in the East Indies and soon it was paying dividends of 50 percent against Shell's 5 percent. Deterding took every advantage to weaken his British rival: 'The inescapable conclusion [decided Samuel's biographer] is that Deterding was the architect of the Shell's ruin.'[4] As the world oil trade slumped from 1903 Standard continued to cut prices and Shell's tankers lay idle. Standard was moving further into Europe, putting up a factory in Rumania, and Shell was ousted from Germany by the intrigues of its partner, the Deutsche Bank. Finally Sir Marcus was forced to discuss a total merger with Deterding on humiliating terms; not even fifty-fifty, but sixty-forty with Deterding as the managing director. In April 1906 the new giant came into being, with the name of Royal Dutch Shell. The British public and most of the rest of the world except Holland still knew it as Shell and regarded it as a British company. But the British government now regarded it as being Dutch, and hence vulnerable to influence from Germans—a suspicion of which Sir Marcus became painfully aware.

Sir Marcus, behind his civic grandeur, had been desperately anxious for Shell to be accepted as part of the imperial service. Shell from the beginning raised the question that was to run through each company over the following decades; was it an international business, buying and selling oil between any nations, wherever the best profit was to be made? Or was it part of the special interest of a single nation? The general question was brought to a head by the very specific problem of the Royal Navy.

OIL AND THE NAVY

Ever since 1899 Samuel had been campaigning for the Navy to switch from coal to oil. This would suit Shell, because its oil in Borneo was more suitable for fuel oil, and the proposal was also backed by solid research. The Admirals doughtily resisted, partly with absurd pseudo-scientific objections, but also with more reasonable worries: that their safe supplies of British coal would be replaced by dubious supplies from foreigners in distant outposts. No oil had as yet been discovered within the British Empire, and the Admirals were supicious both of Sir Marcus and of Shell, which derived much of its profits from Japan, and which depended

heavily either on a French contract for Russian oil, or on oil in Dutch territory.

Samuel did manage to gain the support and friendship of the sea-dog who was later to revolutionise the Royal Navy, Admiral Fisher. He soon became known as the 'oil maniac', and began urging Samuel to extend his resources to the other side of the world (as he shortly did in Texas). Encouraged by Fisher, Samuel continued his campaign, taking naval officers on voyages on Shell ships, and even offered (in June 1902) to put Government representatives on the board of Shell to guarantee its concern for the national interest. This suggestion was later taken up by Fisher, who saw himself as a possible government director. (Whether he was financially disinterested in his enthusiasm for Shell has never been entirely clear.) Samuel had a chance to prove his point when the warship HMS *Hannibal*, carrying oil supplied by Shell, was given trials in Portsmouth, watched by Admiralty experts and Samuel. But the *Hannibal* was equipped with obsolete burners and as the ship switched from coal to oil, it let out a thick cloud of black smoke, covering the decks with soot. It was a fiasco.

The Admirals continued to suspect the loyalties of Shell with its foreign network. When oil was at last discovered within the British Empire—in Burma—the India Office specifically excluded Shell from concessions (even though Shell had developed the Indian market) on the grounds that it might fall into the hands of foreigners, notably Standard Oil. Thus the Burma oilfields were left to be developed by a new and soon very profitable company, Burmah Oil, owned by a syndicate of Scotsmen. When in 1910 the Navy at last began converting to oil the big contracts went to Burmah Oil. Shell, now merged with Royal Dutch, was more than ever suspect.

In 1911 came the climactic year when Winston Churchill, then a budding Liberal politician of thirty-seven, was appointed as First Lord of the Admiralty. Fisher was a close friend of Churchill, and urged him to collaborate with Shell and Samuel ('He's a good teapot', he wrote to Churchill, 'though he may be a bad pourer! . . . Old Marcus is always offering me to be a director . . .').[5]

Sir Marcus gave evidence to an Admiralty committee and explained how his company was loyally British, how there was 'no alliance, no agreement, no treaty of any kind between ourselves and the Standard Oil Company'. His evidence was supported less

honestly by Deterding. The commission appeared impressed by
Samuel's case, but soon there was a public outcry against the
rising price of petrol in Britain, which was now becoming—with
the advent of the automobile—a more exposed political question.
The taxis went on strike, there were protests against the 'petrol
ring', and Sir Marcus insisted on proclaiming the blunt truth to
the *Daily Mail*: 'The price of an article is exactly what it will fetch'.
The politicians were reminded that Shell was in business for profit
and that the price of its oil, whether for taxis or for battleships,
would be pushed up by the new demands from the Navy.

Churchill, urged by Fisher, talked to the Shell leaders. He was
distrustful of Samuel but fascinated by Deterding—though it was
Deterding who was the foreigner, and the more ruthless businessman. Fisher insisted that Deterding was becoming rapidly Angli-
cised and urged Churchill to knight him ('He has a son at Rugby
or Eton and bought a big property in Norfolk and is building a
castle! Bind him to the land of his adoption!'). Churchill, however,
was increasingly suspicious of Shell, the more so as the price of oil
fuel went rapidly up, and Samuel forecast further increases.
Churchill, in tune with the anti-monopolist mood of the Liberals,
denounced the price increases, with some over-simplification, as
the result of secret price-rigging between the oil monopolies.[6]

Meanwhile Churchill was being lured by another prospect. He
was eycing with increasing interest the new Anglo-Persian oil
company, later known as BP, which had immensely promising oilfields in the Middle East. After protracted arguments and negotiations a sensational agreement was finalised. Only three months
before the outbreak of war, the British government announced
that it would buy a controlling interest of 51 percent in Anglo-
Persian.

In June, Churchill made a historic speech to parliament explaining his decision with a mixture of jingoism and foresight. He
described how the two giants, Shell and Standard Oil, shared the
world between them (ignoring the fact that Standard Oil had been
broken up three years before) and made much of the power of the
monopolies. 'Against this, amongst British companies who have
maintained an independent existence, the Burmah Oil company,
with its offshoot the Anglo-Persian Oil Company, is almost the
only noticeable feature.' He warned how, in the face of the giant
companies, the buyer of oil may be forced to pay an artificial

price, and he developed what was in effect a devastating attack on Shell:

> It is their policy—what is the good of blinking at it—to acquire control of the sources and means of supply, and then to regulate the production and the market price . . . We have no quarrel with Shell. We have always found them courteous, considerate, ready to oblige, anxious to serve the Admiralty and to promote the interests of the British Navy and the British Empire—at a price.

He beamed at Sir Marcus' brother, Samuel Samuel, MP, as he said it and continued: 'The only difficulty has been price. On that point, of course we have been treated with the full rigour of the game.' The British government would still need oil from Shell, Churchill explained, but 'we shall not run any risk of getting into the hands of these very good people'. The government, he knew, could use the leverage of Anglo-Persian to make sure that it obtained fair prices from Shell.

Samuel smarted under Churchill's innuendoes against Shell's loyalties and business methods, which some people suspected were tinged with anti-semitism.[7] At the next annual meeting he once again protested his company's patriotism. In fact after the World War broke out six weeks later, Shell were able to prove their concern for the British national interest: Deterding made clear that though Holland was technically neutral, the company was firmly on the side of the allies. Shell, though making vast profits from its Far Eastern trade, did not take advantage of the shortage of tankers to overcharge the Navy, which relied more heavily on Shell oil than on Anglo-Persian.

Sir Marcus, after his many slights and rebuffs, retired into honourable old age, bought twenty acres of Mayfair and was created Lord Bearsted. But he never received quite the recognition he had sought as a national hero. Partly no doubt this was because he had lost control of his company to a Dutchman, but also because he was up against the recurring dilemmas of the oil industry: can or should a company have any loyalties besides those to its shareholders? To whom is it really responsible?

Samuel, with his cosmopolitan experience and his trading genius, stamped his character firmly on Shell. The company has ever since been the most internationally-minded of the seven, and the most flexible, preoccupied with markets more than production. The fleet of ships that he built up—before long a larger fleet than the

Royal Navy—could serve as a symbol of the next phase of the
industry, which was to have no geographical centre or base, with
its heart on the high seas, between a temporary source of supply
and a changeable destination, owing allegiance to no single country.

BRITISH PETROLEUM (BP)

The Anglo-Persian Oil Company, of which Churchill had
bought half in 1914, was to become a new kind of industrial
animal. It grew rapidly first as Anglo-Persian, then as Anglo-
Iranian and now as British Petroleum or BP (by which name I will
call it throughout). It posed the other side of the question raised
by Shell, and still does: what is the advantage to a government in
having its own oil company?

Even before the British Government controlled it the company
had a special sense of mission, and was firmly identified with
British interests. The real founder of the company was William
Knox D'Arcy, an engaging adventurer who went out from London
to Australia and made a fortune in the gold-rush. He retired back
to London, a portly man-about-town with a house in Grosvenor
Square, until in 1901 he heard of a report by a French geologist of
great potential oilfields in Persia. Fired no doubt by the American
oil stories (it was the year of Spindletop) he sent out two men to
investigate a possible concession. They negotiated with the Grand
Vizier in Teheran against some pressure from the Russians and
eventually—taking advantage of the absence of the only Russian
in their embassy who could read Persian—they signed a Concession
covering 480,000 square miles (nearly twice the size of Texas) in
exchange for £20,000 in cash, 20,000 one-pound shares, and
16 percent of the net profits.

To begin the exploration, D'Arcy appointed a self-taught
geologist, George Bernard Reynolds, who must rank with Drake
and Lucas as one of the great pioneers of oil. He was a tough loner
of fifty, awkward, easily taking offence and contemptuous of the
'office wallahs' back in London.[8] He went out to the appalling
heat of Chiah Surkh on the western edge of the concession, and
with a crew of Canadians, Poles and tribesmen continued drilling
for three years with no real success. D'Arcy, having spent over
£200,000 could afford no further loss, particularly since his own
tastes in London were extravagant. He therefore approached
Burmah Oil, which was now a rich and respectable company,
with good Navy contracts. The chairman was Lord Strathcona, the

self-made Scots millionaire who had financed the Canadian Pacific Railway. The British government, already interested in Persian oil, encouraged the union, and Burmah Oil agreed to put up extra capital for D'Arcy's venture. Reynolds and his team shifted their drills to a new territory and after another two years, on May 26, 1908, they finally struck oil, spurting fifty feet out of the barren ground. The transformation of Persia had begun.

A year later the Burmah Oil company and D'Arcy formed the new Anglo-Persian Oil Company with Strathcona, now eighty-eight, as the chairman. D'Arcy was given a million pounds' worth of shares but the largest shareholders continued to be Burmah Oil, who contributed part of their character to the new company. Scots engineers and accountants dominated the headquarters, with a mixture of adventurousness and meanness. (In later years Burmah Oil, after their Burmese oil was exhausted, relied on their 23 percent share of BP for their profits, until in the 'sixties they began hectically buying up other companies, including Castrol in Britain and Signal in America, and gambling heavily in tankers. With the slump of 1974 they crashed and their BP shareholding was taken over by the Bank of England—a tragic end to the oldest British oil company.)

From its beginning in its exposed Middle Eastern position Anglo-Persian enjoyed the special protection of the British government, which in turn dissuaded it from selling out to foreigners.[9] The Indian government sent out a detachment of soldiers to guard the drillers, led by Lieutenant Arnold Wilson (who later became resident director of the company). The British resident, Sir Percy Cox, intervened with the local sheikh to agree to carry the oil to the island of Abadan where it could be shipped by tanker across the world. The 130-mile-long pipeline, the first of the great oil arteries of the Middle East, was laid across the bare mountains to the sea; and a refinery was built at Abadan, which was to become the jewel in the company's empire.

In the meantime, the company was busily lobbying the British government for financial support, a long process which has only recently come to light.[10] Charles Greenway, the managing director, began pressing the Foreign Office for a subsidy in 1912, largely on the grounds that they were threatened by Shell, which (Greenway insisted) was a foreign company. Greenway carefully stressed to the Liberal government the immorality of allowing the Shell marketing monopoly to be extended to production. The British

Foreign Office were convinced, but the Admiralty were very wary
of subsidising one source of supply at the expense of others and of
the commitment that might follow investing in Persia, which might
call for military intervention.

After Churchill arrived in the Admiralty in 1911 he soon became
convinced that he could not be sure of Shell, and the Admirals
were more attracted by the prospect of very cheap oil. At first the
Admiralty envisaged a long-term contract with Anglo-Persian, but
by July 17, 1913, when Churchill spoke to parliament, he had
committed himself to ownership. He insisted that 'safety and
certainty in oil lie in variety alone.' But he also proclaimed that 'we
must become the owners, or at any rate the controllers, at the
source of at least a proportion of the supply of natural oil which
we require'. Churchill and the Admiralty were not, as was often
suspected, persuaded by imperial ambitions, but by the simple
advantages of price.

Churchill sent an oil commission out to Persia which included
Professor John Cadman, the leading British expert on petroleum
technology, who returned to pronounce enthusiastically about the
oilfields. Soon afterwards the Admiralty agreed to pay two million
pounds for 51 percent of the company, whose capital was to be
correspondingly increased. The agreement stipulated that the
Company must always remain an independent British concern, and
that every director must be a British subject. The Government
would appoint two directors, who would have the right of veto, and
they assured the company that the veto would only be exercised
on questions of foreign or military policy, or on matters directly
bearing on Admiralty contracts.

Charles Greenway took over as the new chairman, an increasingly stuffy figure who ruled the company for the next
thirteen years, commemorated by Upton Sinclair (in his novel *Oil!*)
as 'Old Spats and Monocle'. He was not an inspiring leader: when
he first went out to Persia in 1911 he fired the man who had discovered the oil, George Reynolds, who then retreated to Venezuela.
Deterding was probably right when he complained that without
the government's backing, Greenway 'would certainly not have
been anywhere near the head of a corporation'.[11] But the early
years of BP did not call for any special entrepreneurial skill; it had
a captive production, and a captive market, protected by the army
in order to supply the navy.

The commercial success of the company was at that time

difficult to judge. The British government would never reveal the price at which the company sold oil to the Navy, and it was not until fifty years later that the original terms were disclosed. In fact, they were very favourable to the Navy. There was a fixed rate of 30 shillings a ton exclusive of freight and a rebate up to 10 shillings a ton according to surplus profits: the rebate had already reached its maximum by March 1921.[12] By the post-war years, the oil from BP was a good deal cheaper than that from other companies. At the same time BP was also declaring handsome dividends, from which the Treasury collected its half; and Churchill was able to boast of the great financial rewards to the Government.[13]

The Persian oil was a great bargain. Later governments were grateful for Churchill's bold initiative, which seemed the more remarkable coming from a subsequent bitter opponent of nationalisation. In 1923 Baldwin's Conservative government considered selling its share back to Burmah Oil, but soon afterwards the Labour Party came to power, and kept it, and thereafter BP has remained half-nationalised.

The Government stake in BP seemed doubly attractive to Churchill in his Liberal days. Not only did it ensure the Navy's supply, but it provided a competitor to the international cartel of Standard Oil and Shell. Churchill was undoubtedly influenced by Teddy Roosevelt's trust-busting zeal, and the formula of half-nationalisation was in some respects the British equivalent of the anti-trust movement, to limit the power of big business: a policy which other European nations were later to imitate. By buying-in to BP the British government proclaimed that oil was too important to be left to the oil companies—an awareness that was to recur periodically over the next decades.

But the BP solution was not quite as easy as it looked. In the first place, like subsequent nationalised industries, it soon began to show the Frankenstein syndrome, growing out of control. Governments sometimes felt they had the worst of both worlds. Abroad, the company was resented by foreign countries, particularly colonial countries, as an arm of the government. At home the government could not influence its policies; the two government directors—often retired Treasury men or diplomats—were not usually knowledgeable enough to argue with the oilmen, and were happy, as we shall see, to 'go native'.

In the second place, the military case for Government owner-ship was very doubtful, for in wartime the security of oil supplies

depended on command of the seas rather than ownership of the oil. 'If the Anglo-Persian Company had been controlled by Germans', asked two shrewd critics of the industry, Davenport and Cooke, in 1923, 'would not the British Navy have had Persian oil just the same? Undoubtedly, it would'.[14] Government ownership, they complained, meant 'pitting the British State against the nationals of other States, not only in producing but in consuming countries'. They warned, with some foresight, that the British example would encourage nationalistic oil policies among others, particularly the French ('nothing is too ridiculous to happen if ever absurdity in oil principles and French militarism go hand in hand'). The argument was thus already being put forward in favour of multinational corporations—that they serve as buffers to soften confrontations between states, whereas national companies exacerbate the conflicts.

And the Admirals who complained in 1912 that there might be problems of military intervention, for all their dogged conservatism, had some prescience. In spite of its commercial expansion, BP in Persia remained dependent on tacit governmental support— perhaps more dependent than its managers realised. It represented a single easy target for any nationalist movement, and could easily help to provoke one. However much Churchill insisted that 'safety and certainty in oil lie in variety alone', the existence of this single huge source of cheap oil encouraged a growing dependence on one country's supply. The British governments never really faced up to the question of whether they might have to intervene to safeguard it—until it was forced upon them forty years later.

NOTES

1. Address in Atlanta, Georgia, Oct 16, 1953: quoted in Robert Engler: *The Politics of Oil*, Chicago, 1961, p. 367.

2. See Robert Henriques; *Marcus Samuel*, p. 109; to which I am much indebted in this section.

3. Henriques: p. 427.

4. Robert Henriques: *Sir Robert Waley Cohen*, London, 1966, p. 126.

5. Randolph S. Churchill: *Winston S. Churchill*, Companion Vol. 2, London, 1969, p. 1927.

6. *Past and Present*, No. 39, London, 1967, p. 150.

7. Henriques: *Marcus Samuel*.

8. See R. W. Ferrier's article in *BP Shield*, May 1972.

9. Marian Jacks: 'The Purchase of the British Government's shares in the British Petroleum Company, 1912–1914': *Past and Present*, London, p. 141.

10. Ibid: p. 141.

11. Ralph Hewins: *Mr. Five Per Cent*, London, 1957, p. 81.

12. Randolph Churchill: Part 3, 1911–14, p. 1963. See also Jacks, p. 166.

13. Winston Churchill: *The World Crisis*, Vol. 1 (1911–1914) London, 1923, p. 134.

14. E. H. Davenport and Sidney Russell Cooke: *The Oil Trusts and Anglo-American Relations*, London, 1923, p. 57.

The Carve-up

Oilmen are like cats; you can never tell from the sound of them
whether they are fighting or making love.

Calouste Gulbenkian[1]

Oil profits generally seem to find their way by some invisible
pipeline into private pockets.

Lloyd George, March 26, 1920

These seven companies—five American, one British, one Anglo-
Dutch—had all become major powers in the oil industry before
the 'twenties. There were many others, including the other offspring of Rockefeller's Standard Oil, that were to play important
roles. But it was the seven 'majors' who were to dominate the
world oil business in the following decades, and to become new
kinds of industrial organisation—in some respects the forerunners
of the modern multinational corporation. Each of them soon
developed into an 'integrated oil company' controlling not only
its own production, but also transportation, distribution and
marketing. With their own fleets of tankers, they could soon operate across the world in every sector of the industry, from the 'upstream' business of drilling and producing at the oilfields, to the
'downstream' activity of distributing and selling at the pumps or
the factories. And each company strove, with varying success, to
be self-sufficient at both ends, so that their oil could flow into
their tankers through their refineries to their filling-stations. It was
this world-wide integration, together with their size, which was the
common characteristic of these seven.

There were plenty of signs that the seven companies were competing, often ferociously, to sell their precious fuel: and nowhere
was the competition more evident than in the promotion of the
new product gasoline for the new automobiles. The names
Standard, Gulf or Texaco, first on cans, then on filling-stations,
then on the bright signs sticking up from the landscape were
visible symbols of the choice facing the consumer. Yet the com-

when the fight was most critical, the governments were out of the ring. The companies—as we shall see—usually had their own way in the end. But it was never quite clear who was using whom.

To radical critics, it looked as if the State Department had simply abdicated the whole process of oil diplomacy to the oilmen. The government, however much they might distrust the oilmen, were not prepared to set up their own organisation. They preferred to use the oil companies, at a discreet distance, as the instruments of national security and foreign policy. The British government, being totally dependent on imported oil, had closer relations, as we have seen, with the two oil companies. But they too preferred not to be closely involved with such a controversial commodity, and, in Britain, too, the heads of the companies began to play important roles in forming foreign policy.

TEAGLE V. DETERDING

It was the competition between the two biggest companies, Exxon and Shell, which set the pace and the pattern. They were companies with almost opposite origins and characters: the first born of a domestic monopoly, firmly rooted in America; the second, a trading and shipping company intrinsically international. But they had resemblances, too. In each company one man dominated for over two decades: Walter Teagle of Exxon, and Henri Deterding of Shell. Both were autocratic and powerful characters, both spent their spare time shooting on their large estates and liked to combine hunting game with hunting markets. Both strove to achieve a world oil monopoly from quite different viewpoints, until they reached a truce, appropriately enough in a castle. It was the Baroque Age of oil.

Walter Teagle came from the oil aristocracy; his maternal grandfather, Maurice Clark, was one of Rockefeller's first colleagues and his father was a partner in one of the earliest Pennsylvania companies. Like Archbold, he first heroically competed with Rockefeller, but then secretly agreed to sell out, while still pretending to compete. As a young man his son Walter found himself taken over by the giant trust. But he still retained an entrepreneurial spirit and became a director of Standard Oil, at the early age of thirty-three. After the dissolution in 1911 he became a director of the rump company Exxon and an increasingly dominant influence, resenting the legalistic caution of his elders.

When Archbold retired, after a brief interregnum Teagle was

elected President and remained there for twenty years. He stipulated that he must have three months' holiday a year to pursue his
hunting, shooting and fishing, and the board's acquiescence
emphasised his authority and his isolation. A tall, thick-set man,
with a boyish smile, he liked to do business on his estates; he
defied the lawyers, and refused to have one on the board. He had
his own private spy, Flanagan, who roamed the world reporting
only to him.[5] He mastered the facts of the world's oil and blud-
geoned his opponents with them; he had no inhibitions about
using his company's financial power. Politically and diplomatically
he was naive; he had inherited a blunt instrument, and used it
bluntly with an assurance fortified by his life-style.

But his aggression was also partly the product of his company's
predicament. Exxon at that time saw itself not as a confident top
dog, but as a vulnerable underdog, a dismembered body, with a
dangerous imbalance between its big markets and its tiny supplies
of its own crude oil. With the post-war shortage of oil it felt a
desperate need to become a fully integrated company, so that it had
quickly to find oil, or to buy other oil companies. In 1919 Teagle
made his first bold domestic move and bought (secretly, in the
Rockefeller tradition) a prosperous Texan company misleadingly
called Humble, which had originated at Spindletop. But he was
still hungry for oil and was preparing to adventure abroad.

His chief rival for the world's oil was Henri Deterding, a much
bolder leader; and the position of Shell, with no supplies at home,
called for greater daring. Deterding was one of those accountants
of genius who see balance sheets in terms of military campaigns;
and Lord Fisher appropriately called him the Napoleon of oil.
From his boyhood in a bank (as he later explained) he discovered
his capacity to unravel complexities, and to make complicated
problems appear simple.[6] He was a small man, with no formal
education and a bad temper. But he had charm, piercing eyes and
a driving ambition. His Dutch colleagues called him by his initials,
HAW, which in Dutch stands for 'He knows everything'.

He never belonged anywhere—his first ambition had been to
be a sea-captain like his father—and he identified himself with a
succession of different countries. When he became managing
director of the Shell group in 1907 he became quite English,
bought an estate in Norfolk and ran the group from London;
after the war (as Admiral Fisher had recommended earlier) he
was knighted for his services to the Empire. But as Shell's am-

petition also had striking limitations. It was not just that the prod-
duct, as far as any consumer could detect, was identical—much more
identical than cars or soaps. Nor that the new garages and filling-
stations seemed to huddle in clusters on the roadside as if they
dreaded to stick their tall necks out alone. More seriously disturbing for the advocates of free enterprise was the tendency of the
giant companies, as they ventured further abroad, to cling together
in consortia and to reach hidden understandings with each other
in their attempts to bring order to the volatile market. The name
the Seven Sisters, that came to be applied to them and which they
so much resented, was not altogether inappropriate. Like the
classical sisters, who were translated by Zeus into stars, they
seemed to have acquired immortality. But also like mortal sisters,
they fought and competed with each other, while still preserving a
family likeness and closing ranks when challenged by outsiders.[2]

The American companies had begun with their production
safely based in their own home country, and their government was
more concerned with protecting the consumer than with national
security. But as the future of American oil became more uncertain,
the companies began to look abroad for their supplies and thus to
become much more deeply involved with diplomacy. The relationships between the company headquarters and Washington thus became both closer and more barbed.

The First World War made all Western governments painfully
aware of the importance of oil for survival, and led to the development of what was called 'oleaginous diplomacy'. The dependence
on oil for survival became obvious as the war extended—fought
with planes, cars, and tanks—and the oil tankers were critical for
supplies. 'Oil,' said Clemenceau, 'is as necessary as blood.' 'We
must have oil,' said Foch, 'or we shall lose the war.' 'The allies,'
said Lord Curzon in a phrase much quoted by oilmen, 'floated to
victory on a wave of oil.' The Germans were desperately short of
oil, while Britain had access to the Persian oil from BP at Abadan,
and to Shell oil from Mexico and the East Indies. But still by far
the biggest source of supply was America and 80 percent of all
allied oil supplies came from the United States. A quarter of all
the allies' oil came from a single company: Exxon.[3]

After the war the United States was still easily the leading oil
producer in the world and far the biggest consumer. In the post-
war years there was a new rush of consumption, with the multiplication of automobiles and the building of a new life-style based

on cheap oil. The right to travel cheaply, to have cheap electricity and cheap heating, became regarded as part of American democracy and the whole landscape was already being transformed by the product. As the two English authors Davenport and Cooke described it in 1923:

> Travel but a little in the country and you will gain the impression that the modernism of the United States flowed from its oil wells. Outwardly, oil occupies there the place which coal occupies in Great Britain. The oilfield derrick is as familiar a landmark to the American as the pithead wheels to the workers in our 'black country'. The oil-tank car is as ubiquitous on his railroads as the coal-truck on ours. The oil-tin litters his waste places. His wayside is dotted with the petrol pump and at night illuminated oil 'filling stations' make his streets beautiful. A network of oil pipe-lines underlies his country, more extensive than the network of railways overlying ours ... Does not the American partly live in oil? Certainly he cannot move without it. Every tenth man owns an automobile, and the rest are saving up to buy one.[4]

But now there was once again the nightmare of a world shortage. The American Secretary for War, Josephus Daniels, who had confronted Socal over the Californian naval reserves, was appalled by the rapid depletion; and in 1920 the Director of the U.S. Geological Survey pronounced that the American oil situation 'can best be characterised as precarious'.

Both Americans and Europeans began a new scramble for oil, with new bitterness between the haves and the have-nots. The British and French tried ruthlessly to establish their own sources of oil in the Middle East, which they regarded as their own equivalent of Texas. They were determined to exclude America, which had its own oil anyway. But the Americans argued that it was quite wrong that they should provide the world out of their dwindling reserves, while they were shut out of foreign oil. They had helped to win the war, and were entitled to its spoils. In the post-war years, both companies and governments came into sharper conflict. The State Department supported more closely the interests of oil companies abroad, and the new intimacy was further encouraged by the administration of President Harding (whose benign attitude to oilmen eventually extended into the Teapot Dome Scandal). The relationships were often confusing. Ostensibly the companies were the boxers in the big fights, and the governments were the seconds, providing encouragement or reproof. This meant that

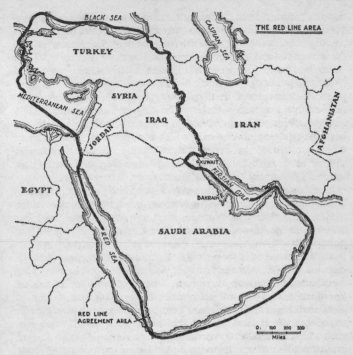

it President Poincaré set up a national oil company, partly modelled on BP, called the Compagnie Française de Pétroles (CFP), with a distinguished scientist and *Polytechnicien*, Ernest Mercier, at its head. The French government gave it special protection and built refineries for the oil from Iraq. But the CFP did not have the same scale or range of production as the others and the French resentment at the Anglo-American domination of the industry was to smoulder, with periodic explosions, over the following decades.

THE RUSSIANS

While Teagle had established himself without too much exertion in the Middle East, he had been making a much bolder and rasher bid for oilfields in Russia. Russia's huge reserves had ever since the 1880s provided alternately threats and temptations to the American companies, as they would do for another fifty years.

Exxon's dealings with the revolutionary Soviet regime revealed a good deal of the political naivety and arrogance of the oilmen, and conjured up for the first time the spectre which was later to haunt the industry round the world—nationalization.

At the time of the Revolution in 1917 Russia was producing about 15 percent of the world's oil; a third of her oil was controlled by the Nobel Brothers, and Shell controlled some of the rest. After the October Revolution there was chaos and Bolsheviks rebelled in the oilfields. Oilmen rushed to the scene, hoping to pick up bargains from the confusion, and Teagle was approached by the Nobel Brothers (who desperately wanted American diplomatic support) to buy half their interest. While Exxon and the Nobels were still discussing terms, the Red Armies seized all the oil properties, but astonishingly, Exxon continued negotiating: as Teagle calmly explained, there was 'no alternative but for us to accept the risk and make the investment at this time'. After consulting John D. Rockefeller II and the State Department (who would not promise diplomatic support) Teagle made an agreement in July 1920 to buy out the Nobels with $11·5 million dollars paid secretly through a Swiss company to disguise Exxon's ownership. The agreement, the company historians admit, 'appears to have been one of the most ill-considered ever made' by the company.

While Teagle and the Exxon board waited expectantly for the collapse of the Bolsheviks, the Red Army remained. Teagle was soon in the quandary that would so often repeat itself in the oil industry. On the one hand he was righteously indignant about the Communists' expropriation, 'the one fundamental thing about Communists that we need to know, and that so many generally overlook, is that the Communists repudiate the civilised code of ethics ...'[10] On the other hand the Russians were now busily producing oil, offering it at cut prices, and Exxon's competitors including Shell and BP were trying to make deals with the Soviets. The nightmare loomed of Exxon's European markets being flooded with cheap oil as Exxon had flooded Shell's markets in the past. Eventually in July 1922 Teagle and Deterding with representatives of fourteen other oil companies met in London to try to form a *Front Uni* to insist on compensation from the Soviets. But Teagle quite rightly did not trust Deterding, and just after the policy was published, Deterding began secretly to buy large quantities of Russian oil, which he sold in the Far East to undercut another of the sisters, Mobil.

bitions, and his own, extended he became again rootless and unstable, changing houses, wives and political views until he ended
up living in Germany, a committed Nazi.

Shell was an amazing mixture of bureaucracy and high gambling.
Its managers were a solid respectable elite, a new breed of carefully selected para-diplomats; as early as 1910 Shell had helped to
set up the Cambridge University Appointments Board, to be sure
of recruiting the cream of graduates, competing with the Foreign
Office and civil service for brains. Yet at the top of this hierarchy
was a tycoon of explosive unpredictability, gambling with oilfields,
paying himself a million dollars a year.

Deterding and Teagle fascinated each other. The acquaintance
began at a famous encounter in 1907 when Deterding had just
taken over the merged Royal Dutch Shell and came to see Standard
Oil in New York to try to persuade them to abandon their price-
cutting wars. Deterding realised that price-cutting was not true
competition at all but 'throat-cutting', and he tried to persuade the
directors of Standard including Archbold and young Teagle to
agree to stable prices. Deterding's precept inherited from his old
boss Kessler was the Dutch proverb *Eendracht maakt macht*—
co-operation gives power—and he was always searching for a
global agreement or cartel. Deterding put his case but without
success, for Archbold was determined on price-wars, to extend
Standard Oil's power. Teagle, however, was much more impressed by Deterding's arguments, and Deterding thought Teagle
a 'brilliant young man'.[7] From then on, the two budding tycoons
eyed each other with mutual and wary respect.

Shell had always been forced to search out oil abroad, and
Deterding was prepared to take large gambles. In 1908 he bought
oilfields in Egypt, which for a short time produced handsomely.
He began exploring in Sarawak and in 1910 started new dealings
with Russia, which led to him buying the Russian Ural-Caspian
oilfields. In the next years he pulled off two of his biggest coups,
in countries which Exxon had avoided. In Mexico he bought the
oilfields belonging to Lord Cowdray (see page 100), which soon
pushed up Shell's total production. In Venezuela he bought the
fields which still provide a sixth of Shell's supplies. At the same
time he was starting to challenge the Americans on their home
ground, by buying oilfields in California. Soon the Shell gasoline
signs—the 'yellow peril' it was then called— were beginning to
shoot up across America to the growing alarm of Teagle and Exxon.

Teagle was already alerted by an indiscretion of Shell. In 1919 a public relations man and sharepusher for Shell's Venezuelan company, Sir Edward Edgar, made a historic blunder by accusing the Americans of having recklessly squandered their legacy of oil while the British position, he said, was impregnable. 'All the known oilfields, outside the United States itself,' boasted Sir Edward, 'are in British hands or under British management or control or financed by British capital . . . We hold in our hands the secure control of the future of the world's oil supply.' The rash boast of the baronet was to be endlessly quoted by the Americans; and it helped to goad Exxon to stir abroad.

IRAQ

Teagle was gradually becoming aware of the importance of Middle East oil backed up by the State Department, in the face of the post-war panic about the shortage of oil. To encourage the companies Washington proclaimed the doctrine of the 'Open Door' which demanded that the Allies should not discriminate against each other in oil supplies. It was a plausible liberal policy, but over the following years the Open Door proved to be a mysterious portal, with a habit of swinging shut again, just as the Americans had got inside.

The immediate battlefield for post-war oil diplomacy was the disintegrating Ottoman Empire. Turkey was paying for defeat by having her dwindling possessions carved up between Britain and France. Both countries, while pretending that oil was not foremost in their minds, were specially concerned with two regions along the river Tigris in Mesopotamia (soon to become Iraq), the regions of Baghdad and Mosul which were suspected of containing huge oil reserves. The real originator of this interest was neither a company nor a government, but the extraordinary Armenian entrepreneur, Calouste Gulbenkian, the first of the lone operators who were to enrich themselves from the Middle East. Gulbenkian as a studious young graduate in Constantinople was always convinced that there was oil along the Tigris, and in 1914 he had put together a syndicate which was then called the Turkish Petroleum Company (TPC). Almost half was owned by BP, who already had the concession in Iran next door; about a quarter by Shell and a quarter by the German Deutsche Bank, who had financed the Baghdad railway. Five percent—the famous five percent— went to Gulbenkian. The agreement went into abeyance during the

war, but in December 1919 it was revised at the conference at San Remo, convened to draw up the peace treaty with Turkey. It divided the Arab countries which had been under Turkish rule into British and French Mandates, and it added a special oil agreement revising the old Turkish Petroleum Company to give the German quarter-share to France instead. But Gulbenkian kept his five percent.

It was a classical European horse-trade, and it deliberately excluded the United States, on the semi-plausible grounds that America had not declared war on Turkey and was not therefore concerned with the peace treaty. But the Americans were outraged when the agreement came to light. The American ambassador in London delivered a strong note to the Foreign Office implying that Britain was trying to corner the world's oil, and recalling (in stately language) that America had helped to win the war and was entitled to a share in the spoils. Lord Curzon, the British Foreign Secretary, replied that oil from the British Empire and Persia amounted to 4·5 percent of the world's production, whereas the United States controlled (with Mexico) about 82 percent.

By August 1922, under sustained pressure from Washington, the British offered the Americans first 12 percent of the Turkish Petroleum Company, and eventually 20 percent, which they accepted. The State Department had gradually pushed the door into Iraq open, but the oil companies went through it with some reluctance. Exxon led the way and eventually seven companies, including Gulf, Texaco and Mobil, agreed to go into Iraq (though all but Exxon and Mobil eventually dropped out). This first venture of American oilmen into the Middle East was not exactly fraught with enthusiasm and enterprise. As the Gulf representative, Charles Hamilton, described it, 'representatives of the industry were called to Washington and told to go out and get it.'[8]

There followed further obstacles, including a war between Greece and Turkey. Curzon, while hardly mentioning the squalid word oil, came near to threatening war over the issue; and eventually Mosul was safely included in Iraq under the British mandate. By 1925 the new government of Iraq reluctantly signed an agreement with the old Turkish Petroleum Company, soon to be renamed the Iraq Petroleum Company, granting a concession until the year 2,000. It decreed that the company must remain British, and its chairman a British subject, and it promised the Iraqis a

royalty of four gold shillings a ton. In the original San Remo agreement there had been an important stipulation that the Iraq state
should have a 20 percent participation in the concession—an idea
which proved to be forty years ahead of its time. But this was not
included in the new agreement, which caused the resignation of
two Iraqi ministers in protest. The refusal of the company to allow
participation—except in the form of royalties—was a continual
cause of bitterness with the Iraqis for the next four decades.[9]

Gulbenkian still insisted not only on his 5 percent, but also on
retaining the original clause in which each partner undertook not
to seek concessions in the former Ottoman Empire—stretching
from Turkey down through modern Jordan, Syria and Iraq, down
to the bottom tip of Saudi Arabia—except through the company.
That was, of course, as the State Department pointed out, completely contrary to any principle of the Open Door. But the American companies were now anxious to compromise and after three
years of negotiations all the partners finally agreed in July 1928 at
Ostend in Belgium.

Since no-one was quite sure what *was* the former Ottoman
Empire, Gulbenkian then and there drew a line on the map with
a red pencil round the huge space which *he* meant, including all
the great future oil-producing areas of the Middle East except
Iran and Kuwait. Thus came into being the most remarkable
carve-up in oil history, known ever since as the Red Line Agreement. But no-one, even including Gulbenkian, foresaw the full
consequences of it. The five American companies, led by Exxon,
were now safely inside sharing 23·7 percent of the new company.
The door slammed behind them. It was the pattern of many
future consortia. As Gulbenkian later described it: 'never was the
open door so hermetically sealed'. Six months after drilling had
begun in April 1927, as Gulbenkian had for so long predicted,
there came the discovery of one of the most fabulous oilfields in
history.

The Iraq Petroleum Company was to be both the parent and
the prototype for other joint ventures in the Middle East. The fact
that it included four of the major companies—two British and
two American—made the process of control of production much
easier, and limited the nature of the competition. Also, coming
on top of the British monopoly in Iran, it increased the resentment of other countries without oil. The French, it was true, had
been allowed their 24 percent share of the Iraq oil, and to exploit

23. W. L. Mellon: *Judge Mellon's Sons*.

24. Craig Thompson: *Since Spindletop*, Pittsburgh, 1951, p. 18.

25. Oliver St John Gogarty: *As I was going down Sackville Street*, London, 1937, p. 133.

26. John O. King: *Joseph Stephen Cullinan*, Nashville, 1970, pp. 184–93.

27. King: p. 212.

Teagle, as he made clear in worried memos, was increasingly torn. Russian oil began flowing into Europe. The more other companies made deals with Moscow, the less effective were Exxon's demands for compensation. The diplomatic atmosphere became more conciliatory after Britain recognized Soviet Russia in February 1924. At last Teagle and most of the board came round to the reality. Teagle met again with Deterding, and they agreed to form a joint company in Liechtenstein to seek a long term Soviet contract 'for the restoration of the Russian oil industry'. For the next two years they went on arguing with the Russians, who at one time even promised some compensation. But eventually the talks were broken off by Teagle's insistence on more money, and by sudden indignation from Deterding: 'I feel that everybody will regret at some time', he burst out in January 1927, 'that we have anything to do with these robbers.'

The explanation for Deterding's turnabout was very personal. Deterding had just been re-married, to a forceful White Russian, Lydia Pavlovna Kordoyaroff. Teagle realised that she had a hold on Sir Henri, who had now become piously and militantly anti-Communist. But he had also become furious because Mobil, in retaliation, was now undercutting Shell in India, the soft under-belly of the industry. He launched a fierce press campaign, accusing all the Standard Oil companies of collaborating with Communism, and demanding that Teagle should bring the other companies into line. Exxon in return made a statement implying with less than total candour that they had had no dealings with Moscow. Deterding insisted that he had been double-crossed by the Standard companies and promptly launched a price-war of the kind he had so long denounced, with particular intensity in India.

The Russians, in fact, had succeeded in generating chaos in the Western oil industry, luring one company after another and setting them against each other, and making nonsense of any legalistic or ideological stand. And Teagle's boldest venture in oil production outside America had ended not just in total fiasco, but in what everyone most dreaded, an uncontrolled glut.

THE CASTLE

The price war that Deterding unleashed in India soon spread through the world, and what began as a quarrel about Russian oil developed into a major crisis for the oil industry, forcing some small companies out of business, and reducing everyone's profits. Deterding was now, from an apparent position of strength, playing Rockefeller's old game. But nobody could win, for there was once again a surplus of oil in the world, contrary to all the predictions of seven years before. The great surge in consumption by the new motorcars had begun to abate. There was new production in Venezuela and Mexico, on top of the flood of Russian oil: and in the Middle East the new oil in Iraq added to the flow from BP's oilfields in Iran. In the battle for the world oil there were as yet only three major contenders, Exxon, Shell and BP, and they each dreaded that the price-wars would ruin them all. Deterding's original argument with Standard Oil, about the dangers of all-out competition, now looked much more convincing to Teagle and the Americans.

BP, the junior of the three powers, had become increasingly important on the world's stage because it controlled not only Iran, but also a quarter of the oil from Iraq; it was now much more than an adjunct to the British Navy. Lord Greenway ('Old Spats and Monocle') had given way as chairman in 1927 to Sir John Cadman, the Professor of Petroleum who had advised the company at its inception. He was the first figure of real magnitude in the British oil industry: a tall, commanding presence with wide-ranging interests and a human touch (he was even an amateur conjuror). He dominated the company with Christian paternalism, and though he was always close to British governments he was determined to assert BP's independence. 'The operations of the company have been very successful,' he told the American Petroleum Institute back in 1921, 'but these have been achieved not because of, but in spite of, the Government holding.'[11] It was a very questionable claim.

Early in 1928 these Big Three began to hold secret conferences, initiated by Teagle, to try to limit their destructive competition.[12] In August the newspapers got wind of a mysterious meeting in Achnacarry Castle, a romantic turreted mansion in Inverness-shire in the Highlands of Scotland. The castle was the seat of the Cameron of Lochiel, but the Cameron, it turned out, had rented it to Sir Henri Deterding for the beginning of the grouse-shooting season and for brown-trout fishing in the private loch below. The castle, reported the *Sunday Express*, was now 'an impenetrable fortress which harbours one of the most interesting groups of silent personalities in the world'. It was indeed a strange gathering.

Deterding had with him his new Russian wife and two mischievous teenage nieces. They were joined first by Sir John Cadman of BP and his wife and then by Walter Teagle of Exxon, who had sailed over from New York, trying to remain incognito. (In London, when a reporter had rung up to ask about Teagle's whereabouts, he picked up the telephone himself and replied: 'Teagle? Never heard of him.') Teagle brought his advisers, Moffett and Mowinckel from New York, and Riedemann from Berlin; and representatives of other companies also arrived, including William Larimer Mellon of Gulf. The press failed to penetrate the fortress, and the *Express* reported that 'the Scottish lochs themselves are no more communicative than the oil lords'.

Ostensibly the house-party had assembled for the grouse-shooting, but as Teagle admitted afterwards, 'the problem of the world's petroleum industry naturally came in for a great deal of discussion'. The party was not a great success; it was a queer bunch of people, Teagle complained later. Cadman was a poor fly fisherman, and the hunting was 'lousy'. Lady Deterding's nieces played tricks on Riedemann, putting molasses in his bed, and tying up his pyjamas in knots: 'Riedemann was mad—oh boy, he was mad.'[13] But in their two weeks at the castle the oil barons reached a basic agreement to divide the world into an international cartel. By early September the three leaders, Teagle, Cadman and Deterding, had approved a statement of principles called the 'Pool Association', or as it came to be known later, the Achnacarry Agreement, or more simply, 'As Is'. It was never fully implemented: but the intentions and principles, and the subsequent political reaction, were to re-echo through the industry for the next thirty years.

The secret agreement (not fully revealed until 1952) began by explaining that 'excessive competition has resulted in the tremendous overproduction of today', and the resulting price-cutting had been destructive rather than constructive. It followed with seven principles to govern the proposed collaboration, beginning with the basic rule: 'The acceptance by the units of their present volume of business and their proportion of any future increase in production.' Principle No 3 insisted that 'only such facilities [are] to be added as are necessary to supply the public with its increased requirements of petroleum products in the most efficient manner'.

It was the kind of oil cartel for which Deterding had aimed twenty years earlier when he urged Archbold and Teagle to follow

the Dutch precept of 'co-operation means power'. Deterding and
Shell were now strong enough to prove to Exxon that they could
never achieve their own global monopoly. In terms of oil economics, there was a great deal to be said for it, and Deterding's plan
was in the familiar tradition of European trusts. But in democratic terms, it was intolerable: hence the secrecy. For it abdicated
to a handful of businessmen the right to allocate the oil trade, and
to fix prices.

The cartel suited Teagle well enough, faced with diminishing
profits and a new nightmare of glut. For the basis of the cartel was
to be the maintenance of American prices. The arbiter of the
world price of oil was the 'Gulf Plus System'. To protect American
oil, the oil from anywhere else was fixed at the price in the Gulf of
Mexico, whence most United States oil was shipped overseas, *plus*
the standard freight charges for shipping the oil from the Gulf to
its market. While the United States was the only major producing
country the system had some justification; but as American oil
was becoming more expensive, and threatened by much cheaper
oil from Iran or Venezuela, it was a blatant device to keep up
prices. To take one example provided by an addition to the 'As Is'
agreement: if BP supplied cheap oil from Iran to Italy, the oil
would be charged as if it had come from the Gulf of Mexico to
Italy; and the saving would make a large profit for the company.
Moreover the 'As Is' now proposed that if two companies agreed
to swap production to reduce costs of transport, they would share
the profits between them. This device of 'phantom freight', as it
was romantically called, was already well-established before Achnacarry, but it was now formalised and increasingly beneficial to
both sides of the Atlantic. It protected expensive American oil and
ensured high profits for Shell and BP who were dealing with
cheap oil elsewhere.

The 'As Is' was not so much a list of rules, as a constitution, or
a declaration of intent. It was approved first by the Big Three,
then by fifteen other American companies, including the other
four sisters—Gulf, Socal, Texaco and Mobil. It could not be
totally exclusive, for it could not discipline Russian oil, but it
gave the biggest Western companies powerful advantages. To
supervise the quotas in 'As Is', the companies agreed to set up
two export associations, to protect the members like Rockefeller's
original railway rebates, enlarged to global proportions.

CONSERVATION AND CARTEL

The restrictions of 'As Is' were blatantly designed to maintain profits: but they could also be presented, as Rockefeller presented his restrictions, as being devised to prevent waste, and working under the new banner of Conservation. Not for the last time conservationists and cartelists were in alliance. Within the United States the hectic and disorganised drilling for new oil had often used up reserves prematurely, which had led first to an ineffective Federal Oil Conservation Board in 1924 and then to a more serious effort. By March 1929 the American Petroleum Institute approved a control plan to hold back the production of oil in the United States to the levels of 1928. This conservation agreement neatly complemented the 'As Is' Agreement of six months before, by limiting exports from the United States. In fact, the U.S. Government later insisted that the companies had no legal authority to regulate production, so instead the oil-producing states eventually worked out an 'interstate compact' by which they would jointly restrict oil production.

There was always a useful confusion about whether this restraint was to prevent wasteful drilling, or to keep up prices. When Bill Farish, then chairman of the executive committee of Exxon (later President), was subsequently asked in Congressional hearings, he replied: 'The underlying thought covered is the whole idea of conservation and stabilisation of the industry, which to my mind are synonymous terms . . .'[14]

It was in Texas that the problem of restricting production came to a head and was eventually solved: and this development, though local, had profound consequences and lessons for the rest of the world. For Texas was the wild edge of the industry; and if it could be controlled, the way was open for stabilisation elsewhere. The crisis began near the small town of Kilgore in East Texas, where a wiry old wildcatter called 'Dad Joiner' was persisting in drilling on a farm belonging to a genial widow called Daisy Bradford, in a bleak territory which had been written off by the big oil companies and their geologists. In October 1930, Dad Joiner drilled further, and black oil spurted out at the rate of 7,000 barrels a day. It set off, in the age of the motor-car, a more ruthless oil-rush than Titusville or Spindletop: in Kilgore, the oil derricks overlapped each other on backyards, front gardens and churchyards; and the strike turned out to be only part of a huge

East Texan oilfield. Dad Joiner himself suffered the usual pioneer's fate: eventually he retired in penury to a small house in Dallas, in Mockingbird Lane. The man who bought his land, H. L. Hunt, grew to be a billionaire, the richest of all the Texans.

The East Texas boom, coming at a time of both depression and overproduction, produced a new glut of oil, when it was least of all welcome to the big companies, and it looked like a return to the wild old days of individualism before the Big Hand of Rockefeller reached out. Drillers rushed to get up the oil before their neighbours, thus quickly reducing the pressure and potential of the oilfields. The problem of control was made harder by the American 'rule of capture' by which oil was regarded in legal terms like a wild animal—who ever caught it first, could keep it. Prices fell down to 10 cents a barrel, and gas-stations competed with free chickens to lure customers.[15]

Eventually the chaos and anarchy called for intervention—not this time by a company, but by the State: in the end the Governors of Texas and Oklahoma called in the national guard to close down oilfields, and enforced a system of rationing, by which the demand in any one month was shared among oil producers by a state body called the Texas Railroad Commission. It was difficult to enforce and many producers secretly sent oil out of the state, where it could be sold freely, until a 'Hot Oil Act' was passed in 1935. But remarkably, Texas did succeed in controlling output, with the reluctant support of the producers; for they now realised that it was the only way to keep up prices. With the state's help, they achieved what the Pennsylvania producers never did. The implementation of 'pro-rationing', as it was awkwardly called, was a milestone in the industry, and its importance was not unnoticed in the following years by the shrewder men in producing countries abroad: the Venezuelans actually took the precaution of hiring a man from the Texas Railroad Commission to advise them. The restricting of oil from Texas, the frontier of free enterprise, was a necessary concomitant of any global agreement to maintain prices. Thirty years later, OPEC was to look back to Texas as their model for controlling Middle East production.

'As Is' was never completely achieved; it was impossible to bring into line all smaller companies, and Russian oil was always bubbling up in odd places. But the Big Three, with local allies, could often make deals in individual countries. One of the most effective was Britain, where Shell and BP collaborated with Exxon to fix prices, and also formed a joint British marketing company called Shell-Mex BP (which remained together until the mid-seventies, when it was dissolved after long and painful negotiations). Another was in Sweden, where the market was carved up between Exxon, Shell, BP and Texaco and two other companies. They met each week in the Shell offices in Stockholm to ensure that no company undercut another—which was not fully revealed until investigated by a Swedish parliamentary committee in 1947.[16]

The big companies held successive meetings in the early 'thirties to try to enforce the cartel, and in 1932 six of the seven sisters met to adopt more flexible quotas. The last and most important meeting was in April 1934, when Exxon, Shell and BP met secretly in London to reformulate the Achnacarry Agreement. Together they produced a Draft Memorandum of Principles, to operate throughout the world except where the law forbade it, in the United States. The Memorandum laid down rules for restricting competition and sharing profits from outsiders, including a system of penalties to be enforced by a special 'London Committee' (London, then as later, was the natural venue for cartel discussions). This agreement was ended, according to Exxon's later evidence, in early 1938; but in Sweden at least, from the evidence uncovered it continued through the war.

Most of the world's oil resources were in the hands of the big companies, and the agreements succeeded in their main object of maintaining stable prices at the American levels, and of limiting competition inside each country. The monopoly of Rockefeller had evolved, with apparent inevitability, into a global cartel.

TRIUMPH AND TREASON

Through the 'twenties and 'thirties, Teagle and Deterding were the 'Titans' of world oil, and each voyage across the Atlantic was surrounded by speculation. After the cartel agreement at the Castle their position seemed still more secure, as they covered the world with secret agreements and played producers against each other. But they were not men of political subtlety; they were both autocrats at heart. As the world began to divide and split apart, they were unaware that their careers were heading for danger.

Teagle had succeeded triumphantly in reviving and stabilising Exxon. The Depression weakened his competitors and in 1932 a new source of oil supplies was virtually handed on a plate. The Standard Oil Company of Indiana, another offshoot of Rockefeller's company, had developed a valuable company called Creole in Venezuela, but could not now make use of the oil, with a glut in its home markets and a shortage of finance. Teagle ignored warnings of possible anti-trust action and bought the Creole concessions for $140 million, It was a bargain, and gave Teagle a new counter against Deterding and Shell, already entrenched in Venezuela. In 1937, his last year as President, Teagle told his shareholders that Exxon's share of world oil production had gone from 2 percent to 11·5 percent. By the end of the decade, half of Exxon's production was outside America, and the company had overtaken Shell to become the world's largest producer—a position it has occupied ever since.[17]

Teagle had become established as the doyen of the industry; his relationship with Republican administrations was close, and there was little trouble from anti-trust. He played poker with President Harding (who once asked him to buy a baseball club). He was friendly with President Coolidge (whom he later asked to run the American Petroleum Institute). And President Hoover, after the great crash, used Teagle as one of his business advisors, Even when Roosevelt came to power, Teagle at first worked quite closely with the New Deal. The problems of overproduction had made oilmen more conscious of the need for government involvement and Teagle became chairman of the Industry Advisory Board. He regularly sent Roosevelt parcels of pheasant and salmon,[18] but he soon became disenchanted with the New Dealers. The Democrats, while collaborating with the captains of industry, were well aware of the public hostility to them, and the political advantages of attacking them. And Teagle remained very vulnerable to public attack—more vulnerable than he realised. Exxon had inherited a great deal of the distrust of the Standard Oil Trust, and Teagle maintained the old tradition of secrecy and occasional mendacity. His global deals were shrouded in camouflage, and kept secret even from his fellow-directors.

It was Hitler who was to be the undoing of Teagle. As part of his network of global agreements, Teagle had made an agreement as early as 1926 with the great German chemical combine, I.G. Farben, for an exchange of patents and research. It seemed sensible and far-sighted at the time: Teagle was impressed by the Germans' progress in extracting oil from coal and in other research which led to the development of synthetic rubber, while the Germans preferred collaboration to competition. As Standard Oil summed it up: 'The I.G. are going to stay out of the oil business and we are going to stay out of the chemical business . . .' After Hitler came to power, the co-operation gave Germany the patents for tetra-ethyl lead, crucial to hundred-octane avation fuel. Exxon in exchange looked to the Germans to develop synthetic rubber and held back the research in America. After the German invasion of Europe, Exxon continued to exchange vital information. There was nothing uniquely sinister about the collaboration; in the words of Thurman Arnold, the crusading new anti-trust chief, 'what these people were trying to do was to look at the war as a transitory phenomenon and at business as a kind of permanent thing'. But Teagle remained naively unaware of the political consequences.

In 1941, the Justice Department, having investigated the company's files, brought two anti-trust suits against Exxon. The first was for conspiring to control oil transportation through pipelines, the second for making restrictive agreements with I. G. Farben. Teagle had by now become chairman, and had been succeeded as chief executive by his old Texan fellow-hunter, Bill Farish. But Teagle, as the man who first made the agreements, was the obvious target for the press. ('The plain fact is that Adolf Hitler has used and is using American citizens and American laws to advance the Nazi cause in this hemisphere', wrote the newspaper *PM*.) Thurman Arnold mounted a ferocious attack on Exxon, charging them with blocking synthetic rubber research, and providing Germany with crucial industrial secrets. The case was overstated; and in fact the Americans had learnt much from the German research. But Teagle and Farish had been secretive and rash in their collaboration and had contravened anti-trust laws. Teagle, distraught and bewildered, tried to justify his own position with an anguished letter to Roosevelt, but the President refused to intervene. The rest of the board preferred to settle with the Justice Department, which modified its charges in return for Exxon's promise to release its patents and pay a fine of $50,000.

The attack then passed to another more publicised forum, a Senate Committee investigating national defence, under Senator Harry Truman. Arnold testified to the Senators about Exxon's failure to pursue artificial rubber (it was just when the Japanese had overrun the Malayan rubber plantations). Farish, tense and angry, tried to explain Exxon's dilemma, while Teagle listened in silence to the charges against him. After the hearing a reporter asked Truman whether he regarded Exxon's agreement with I. G. Farben as treasonable, he replied 'Why, yes, what else is it?'

The attacks eventually spent themselves, but the company never altogether recovered its old confidence. Even John D. Rockefeller II, who had kept studiously aloof from his father's creation, felt compelled briefly to investigate. Exxon sales suffered from the public outcry, and the board was determined to re-organise its public relations, to break down the tradition of secrecy, and to inform itself about its own company. As one director, Wallace Pratt, put it: 'I have come, in short, to rate as indispensable to the successful oilman a thorough knowledge of the corporation he directs.'

Teagle and Farish were both broken by the accusations of treason. Farish was still recovering from the experience eight months later, at Teagle's estate, when he collapsed and died. Teagle lost all his customary confidence, became nervous and fumbling, and resigned before the end of his term at the end of 1942.[19]

For Deterding, the end was more justly ignominious. By the mid-thirties, while he was still head of Shell, his behaviour had already become increasingly autocratic with signs of incipient megalomania; and his memoirs published in 1934 showed signs of this: he spelt the word oil with a capital O, and proclaimed: 'If I were dictator of the world—and please, Mr. Printer, set this in larger type—I WOULD SHOOT ALL IDLERS AT SIGHT.' His influence on the company was erratic and as one Shell veteran recalls: 'Deterding's interventions were like thunderstorms; suddenly flattening a field of wheat, while leaving other fields unscathed.' The stately managers of Shell began to have the worrying impression that their Director-General was going mad, and still worse, going pro-Nazi. His anti-Communism, spurred on by his Russian second wife, had already made him sympathetic to the Nazis. But in 1936, just after he had celebrated his seventieth birthday and his fortieth year with Shell, he married a third time, to a German girl, Charlotte Knaack, who had been his secretary. He was now convinced that the Nazis were the only solution to the Communist menace.[20]

With great embarrassment, his fellow-directors finally eased him out of the post he had occupied for thirty years. He then went to live permanently on his estate in Mecklenberg in Germany, making frequent visits to Holland, to encourage closer relations between the two countries, and becoming intimate with the Nazi leaders. He died six months before the outbreak of war: memorial services were held in all Shell offices in Germany and Hitler and Goering both sent wreaths to the funeral on his estate.[21] After the war there was an empty niche in the entrance-hall of the Shell headquarters in The Hague: it was intended for Deterding.

Hitler was also responsible for the disgrace of a third chief executive among the seven sisters: the autocratic President of Texaco, Torkild Rieber. He was an engaging buccaneer, the protégé of the founder of Texaco, Joe Cullinan, and he had a romantic background. He had first come to Texas as a Norwegian skipper's mate, taking oil from Port Arthur. Four years later Texaco bought the tanker and with it Rieber, who helped organise Texaco's tanker fleet. He eventually rose to the top, and 'Cap' Rieber became a famous Houston figure: a stocky broad-smiling man with a thick Norwegian accent, a sailor's cap and a parting in the middle. Like Deterding, Rieber had great daring: he built a 260-mile pipeline across the Andes to carry Colombian oil to the sea, and he pushed into Saudi Arabia by joining forces with Socal (see next chapter). He had a sailor's internationalism, but without any real political instinct: the world was a market with no barriers or taboos.

Rieber first came into trouble with Washington in 1937 when Texaco tankers taking oil consigned for Belgium mysteriously changed direction to Franco's ports in Spain, in the middle of the Spanish Civil War. Roosevelt was furious at this violation of the neutrality law, and his Attorney-General warned Rieber that Texaco would be indicted for conspiracy, but supplies still continued, through Italy. Texaco altogether sent oil worth $6 million to Franco, on credit, to be paid for after the war: Franco's military position would have been much more perilous without these precious cargoes.[22] Rieber also made contact through Spain with leading Nazis and agreed to supply oil from Colombia to Germany. He continued shipments after the outbreak of the European war in 1939, dodging the British embargo by sending the tankers to neutral ports. He could not get money out of Germany. so he made a barter agreement to be paid with three tankers from Hamburg, which served to bring him closer to the Nazis. He saw Goering in Berlin to clinch the deal, but Goering insisted that in return for the tankers he needed something more—Rieber's diplomatic support. Accordingly, in January 1940, Rieber went to see Roosevelt, to put forward the peace-plan devised by Goering to ensure Britain's surrender. Roosevelt ignored it and advised him to back out of his German connection.

But soon after, Rieber was still deeper involved. In June 1940, at the time of the Fall of France, there arrived in New York Dr. Gerhardt Westrick, the German lawyer who had represented several American companies in Germany including Texaco and ITT. Disguised as a commercial counsellor, he arrived on a special diplomatic mission to dissuade American businessmen from supplying Britain with arms, since (he explained) Britain was close to defeat and German-American relations would soon be all-important. Texaco, through Rieber, paid Westrick's salary, and financed his entertainment, including an office in the Texaco head-quarters at the Chrysler building in Manhattan, and a large house in Scarsdale, where Westrick received influential business men and cultivated good will towards the Nazis.[23]

In the meantime a Texaco man in Germany, Niko Bensmann, a veteran oilman, helped Rieber to obtain delivery of the tankers. Bensmann (without Rieber's knowledge) was a skilled intelligence agent, using his own agent inside Texaco's New York head-quarters to obtain a flow of reports from America: including a very accurate review, prepared by Texaco's economists, of the expanding American aircraft industry which was aiming to produce 50,000 planes.[24]

Eventually the head of British Intelligence in New York, the Canadian millionaire William Stephenson, got wind of Dr. Westricks' true purpose, and revealed the story through the *New York Herald Tribune*. Westrick was declared persona non grata, and he left on a Japanese ship to return to Germany, where he took over command of the ITT companies. Rieber was abruptly discredited. Texaco shares slumped on the stock market, and after an angry stockholders' meeting he was compelled to resign, to be succeeded by his rival, 'Star' Rodgers. Texaco's company history makes no mention of the incident: 'It was all got up by the Jews', one Texaco man told me. But at this time Texaco initiated a rare act of patronage—sponsoring the weekly radio broadcasts of the Metropolitan Opera, uninterrupted by advertising, which have continued ever since.

It is not necessary to see these three scandals as evidence of any special moral turpitude on the part of the oil leaders: they were brigands of their time, trying to extend a greedy international industry across the barriers of war. They were men who did not know when to stop, and there was very little to stop them. But their ruthlessness and autocracy did reveal very sharply the basic uncontrollability of oil, and the ability of the industry to defy national governments.

NATIONALISM AND MEXICO

In the great scramble for world oil, there was not much worry about objections from the producers themselves—at least, outside America. In the Middle East, each new discovery appeared to weaken the position of local governments, in the face of a world glut. In Iran, BP maintained its highly profitable monopoly, even after the coming to power of a more formidable new ruler, Reza Shah; and a new agreement was reached in 1933 extending the concession till 1993. In Iraq the IPC consortium could make vast profits with little resistance from the disorganised government, evading the participation clause. Between 1934 and 1939 Exxon made a profit of about 52 cents a barrel; more than twice what they paid to the government. For an investment of $14 million Exxon's share in the IPC was reckoned by 1937 to be worth around £130 million.[25]

But there was a warning of future troubles in Mexico, the country which had the most bitter experience of both exploitation by the oil companies, and the corruption of governments. The most spectacular fortune in Mexico had been made by an English contractor from Yorkshire called Weetman Pearson. He had come to Mexico at the end of the last century at the invitation of the dictator Porfirio Diaz to build a canal to drain Mexico City. He went on to construct the harbour at Vera Cruz, sanitary systems and electric tramways, but his involvement in oil was to have more destructive effects. He happened to be in Texas just when the Spindletop discovery was made in 1901, and immediately bought concessions in likely land in Mexico. He spent £5 million searching for oil, eventually found huge reserves in 1908, and soon afterwards he formed the Mexican Eagle (Aguila) company. There followed ferocious battles with the local distributors, Waters Pierce, in which Standard Oil had a two-thirds share. But Pearson by now had huge resources, and eventually Standard Oil sold out their share, leaving Pearson as virtual master of Mexico.

There were further continuing battles between British and

American interests, each side accusing the other of fomenting political trouble. In 1913, a new dictator, General Huerta, was declared President. Just at this time Woodrow Wilson became President in Washington, full of idealism, and was determined to displace the dictator. He was convinced that he was financed by Pearson, or Lord Cowdray as he now was, with the backing of the Liberal government in London—for Mexican oil had become crucial for the Royal Navy. 'If Taft had had another four years,' wrote the American Ambassador to the President in 1914, 'Cowdray would have owned Mexico, Ecuador, Colombia . . . with such a grip on the Governments as would have amounted to a mortgage.'[26] There followed acrimonious diplomatic exchanges with the Americans accusing the British government of being paid by Lord Cowdray, and the British accusing the Americans of being 'Standard Oil men'; eventually both governments agreed to oppose the dictator.

Meanwhile Cowdray was pumping the oil out of Mexico as fast as he could. Mexican oil played a key role in the First World War, and by July 1918 Exxon reckoned that Mexico's potential production was greater than actual production in the U.S.[27] Mexican oil made Lord Cowdray into one of the richest men in Britain—the nearest British equivalent to a Rockefeller—and he laid the basis of a financial nexus which remains today the most spectacular private empire in Britain, now headed by the third Lord Cowdray, and including such powerful properties as Lazards Bank, the *Financial Times*, *The Economist*, Longmans and Penguin Books.

In 1919, Cowdray sold out the majority of his company to Deterding of Shell; and there followed further prolonged battles between the British and Americans, revolutions and counter-revolutions and growing demands to nationalise the companies. In 1936 the new government of Lazaro Cardenas reached an impasse when workers in the oil fields went on strike against the appalling conditions in shanties and hovels. A government board ordered the foreign-owned companies to meet the workers' demands: but they refused and put their case to the Mexican Supreme Court, Cardenas retaliated by nationalising the seventeen companies in March 18, 1938: it was proclaimed as a new day of independence, and a monument to the nationalisation was put up in Mexico City at which diplomats were required to place wreaths. The oil companies violently protested. But President Roosevelt, having proclaimed his 'Good Neighbour' policy, was determined not to intervene. He insisted that the 'United States would show no sympathy to rich individuals who [had] obtained large land holdings in Mexico for virtually nothing.'

The euphoria in Mexico did not survive. The Americans, British and Dutch boycotted the nationalised oil, and the American Ambassador, Josephus Daniels (the former Secretary of the Navy who had earlier fought Socal) sadly predicted that the Mexicans would 'drown in their own oil'. The new national oil company, PEMEX, was initially corruptly and incompetently run, without enough experts or engineers: and the Mexicans were eventually forced to pay $130 million compensation for seizing the companies. The boycott was lifted to satisfy the demands of the Second World War; but the big companies, having drained off much of Mexico's resources, then switched their chief attentions to Venezuela where they were on cosy terms with the dictator Gomez. By 1946 Venezuela had become the world's biggest producer outside America.

The Mexican nationalisations appeared self-destructive, as if to prove the point that great companies were essential for the survival of the producers. Yet Mexico nevertheless was an important warning to the companies; that they were politically increasingly vulnerable, and could not necessarily rely on home governments to retaliate. Shell took the warning most seriously, and after the fate of the Eagle they pushed through a drastic scheme to recruit local managers. But the other companies were less concerned, and few oilmen expected that Mexico would be the first in a chain-reaction, spreading to Venezuela and thence to the Middle East, which would eventually unify all the producing countries against the companies.

NOTES

1. Nubar Gulbenkian: *Pantaraxia*, London, 1965, p. 96.

2. The phrase Seven Sisters has been applied in many different contexts, and it was used as early as 1913 in connection with the New Jersey Corporation Acts. It was not until the advent of Enrico Mattei, in the fifties (see Chapter 7) that '*le sette sorelle*' became widely used for the oil companies.

3. G. S. Gibb and E. H. Knowlton: *History of Standard Oil Company (New Jersey) 1911–1927*, New York, 1956, p. 244.

4. E. H. Davenport and S. R. Cooke: *The Oil Trusts and Anglo-American Relations*, London, 1923, p. 68.

5. For these and other details of Teagle's methods, see Bennet H. Wall and George S. Gibb: *Teagle of Jersey Standard*, Tulane University, 1974.

6. Sir Henri Deterding: *An International Oilman*, London, 1934.

7. Sir Henri Deterding: *An International Oilman*, London, 1934, pp. 72–6.

8. Senate Hearings on Petroleum Resources, Washington, June 27–28, 1945.

9. See *The Nationalisation of Iraq Petroleum Company's Operations in Iraq*: INOC Publications, Baghdad, 1974.

10. Wall and Gibb, p. 225.

11. See John Rowland and Lord Cadman: *Ambassador for Oil*, London, 1960, p. 101.

12. Federal Trade Commission: *The International Petroleum Cartel*, Washington, 1952, p. 199.

13. Wall and Gibb: p. 261.

14. See FTC Report, 1952, p. 214.

15. Professor Adelman insists that the East Texas glut was quite untypical of the world market, partly because the chaos was caused by the rule of capture, which was uniquely American. See *The World Petroleum Market*, p. 43.

16. FTC Report: p. 266.

17. H. M. Larson, E. H. Knowlton and C. S. Popple: *History of Standard Oil Company (New Jersey), 1927–1950*, New York, 1971, p. 148.

18. Wall and Gibb, pp. 271–281.

19. Wall and Gibb, p. 297–319.

20. See *Het Koninkryk der Nederlanden in Oorlogstyd*, The Hague, Vol. 1, p. 354.

21. *The Times*, February 11, 1939.

22. See Herbert Feis: *The Spanish Story*, New York, 1948, pp. 269–271; also Luis Bolin: *Spain, the Vital Years*, London, 1967, pp. 224–225; also Hugh Thomas: *The Spanish Civil War* (revised edition), London, 1975.

23. See the biography of Sir William Stephenson by Montgomery Hyde, published in Britain as *The Quiet Canadian*, and in America as *Room 3603*, pp. 71–2; also Anthony Sampson, *The Sovereign State of ITT*, London, 1973; also *New York Times*, August 2, 13, 20, 24, 1940.

24. Ladislas Farago: *The Game of the Foxes*, London, 1971, p. 408.

25. See Benjamin Shwadran: *The Middle East, Oil and the Great Powers* (3rd edition), New York, 1973, p. 262.

26. See Desmond Young: *Member for Mexico*, London 1966, p. 1. The true facts about Cowdray's involvement with Diaz and Huerta are hard to disentangle, and both the British biographies are adulatory.

27. Gibb and Knowlton: p. 88.

5

Jackpot

The Arabia which once trafficked in spices and perfumes, for the service of the gods and dead of the ancient world, has risen at last from her long sleep to serve man and Mammon from the new-found sources of her hidden wealth.

Henry St. John Philby, 1951

The trouble with this country is that you can't win an election without the oil bloc, and you can't govern with it.

Franklin D. Roosevelt

While Teagle and Deterding had been carving up the oil business between them, they had both missed out on the exploration that was to change the whole balance of the world, and to make the Latin-American oil fade into the background. The most massive new reserves were discovered by the remotest of the seven sisters, based on California, in a desert territory in Arabia which few American diplomats had ever seen. But they soon caused unique special relationships, a new struggle for control between government and companies, and an inescapable complication in Washington's foreign policy. America ever afterwards was to be deep in the problems of the Middle East.

The Kingdom of Saudi Arabia was the most impenetrable part of the Middle East, hanging like a great sack below the fertile crescent in the North, with only a few green fringes and oases to interrupt the desert. It was a quarter the size of the United States, and as a country it had only just taken shape. In 1927 King Ibn Saud, the desert warrior from the interior, had subdued his rivals in Mecca and the Hejaz on the Red Sea coast with his puritanical army: and he named the whole territory, from the Persian Gulf to the Red Sea after his own clan, Saudi Arabia—the only country to be named after its ruling family.

The possibility of oil in Arabia had already been investigated as early as 1923 by a rough-hewn New Zealander, Frank Holmes, acting for a London syndicate speculating in concessions, whose progress through the Middle East was full of portent. He had rented a concession, but it soon lapsed, and Holmes moved on to Bahrain, a picturesque group of islands off the coast with a community of traders and pearl-fishers. Here he bought a concession from the Sheikh, and it was here that the oil history was to begin. Bahrain at that time was a British protectorate, an adjunct to the Indian empire; its currency was the rupee. The British were thus well placed to control any oil that might be found there, and after BP had found oil in Iran, the Sheikh promised not to discuss oil concessions without British approval. But BP at that time had quite enough oil in Iran and Iraq, and showed no interest in the prospect of oil in Bahrain. They did not take up an interest in Holmes' concession and Holmes went to the United States to sell Bahrain, with other options.

The American sisters responded with great lack of enterprise. Holmes offered the rights to Exxon for $50,000 but Teagle rejected it in what was later described as 'the billion-dollar error'. At last Holmes approached Gulf who paid $50,000 and sent out their top geologist, Ralph Rhodes—later President of the company—who was soon confident that it would yield oil. But Gulf were unable to go ahead on their own because, as members of the Iraq Petroleum Company, they had only recently undertaken not to explore inside the Red Line—which included Bahrain—without the other members. Gulf tried to interest the others, but the geologists of BP, the chief member, were still insistent that Bahrain would not yield oil, since it lacked the 'Oligocene-Miocene' formation found in Iran and Iraq. Gulf, too, then made a historic blunder: instead of pulling out of the IPC, they offered their Bahrain concession to Socal, who were outside the Red Line Agreement.

Socal was an unlikely participant, a massively conservative company across in San Francisco, and many of the directors were sceptical about exploration abroad. They had spent millions drilling unsuccessfully for oil in Latin America, the Philippines and even Alaska and there was now anyway a glut of oil. But two of their directors, Maurice Lombardi and William Berg, were excited by the report from Bahrain, and they now bought the bargain option from Gulf for $50,000.

Socal still had some difficulties with the British government, which insisted that the Bahrain company must have a majority of British directors. But the State Department intervened, quoting once again the Open Door principle, and the restrictions were modified, allowing Socal to own the concession through a Canadian subsidiary. Socal quickly sent out a geologist and by 1931 they had struck oil. Two years later Bahrain was exporting oil to the world market, to the consternation of the other sisters. Bahrain was never in fact to become a major oil producer, and to this day it remains a centre for trading and banking rather than oil. Its real importance in oil history was as the stepping-stone to the mainland of Saudi Arabia, twenty miles across the water.

In 1930, seven years after Holmes' unsuccessful prospecting visit, King Ibn Saud was in desperate need of money. The King was not (like the Sheikh of Bahrain) under a British protectorate and he was in a position of some independence. One of his principal advisers was Henry St. John Philby, the eccentric Arabist who had left the Colonial Service after a row. He had embraced Islam and become close to the King, with a critical view of the British. Philby, motoring with the King one afternoon, suggested that he could find money by exploiting the country's mineral resources, and that his people were 'like folk asleep on the site of buried treasure, but too lazy or too frightened to dig in search of it'. He quoted his favourite passage from the Koran: 'God changeth not that which is in people unless they change that which is in themselves.'[1]

The King quickly reacted, and as a result Philby arranged a meeting with the American philanthropist, Charles Crane, who in turn arranged for an American geologist, Karl Twitchell, to prospect for minerals. Twitchell, encouraged by what he saw, then tried to interest American oil companies on the King's behalf and once again a bonanza was hawked around. He went first to Texaco, who were not interested. He then approached Exxon. He went to Gulf, who again turned down the offer because of the Red Line Agreement. He was then approached by Socal, who were already turning their attentions across the water from Bahrain to Saudi Arabia: Socal now moved swiftly, and engaged Twitchell as their adviser.[2] Twitchell went back to Saudi Arabia to make a deal with the King. Socal also secretly promised Philby a reward if they got the concession.

After the first find at Bahrain, the prospect of oil in Saudi Arabia looked obviously more likely. Sir John Cadman of BP was vexed at having missed out on Bahrain, and insisted that the IPC syndicate must prospect together in Saudi Arabia. They accord-

ingly sent a negotiator, Stephen Longrigg, a former colonial administrator and expert on Iraq, to make an offer to the King. But
the IPC directors, as Longrigg put it, 'were slow and cautious in
their offers and would speak only of rupees when gold was demanded'. In fact, as Longrigg unwisely confided to Philby, the
IPC were not much interested in extracting oil from Saudi Arabia,
at a time of world glut; they were only concerned to keep out the
Americans, and the most they could offer was £10,000 — not in
gold. But Socal, advised by Philby, eventually came up with an
adequate offer to the King: an immediate loan of £30,000 with
another loan of £20,000 eighteen months later, and annual rent of
£5,000 — all in gold. The historical agreement was eventually
consummated in August 1933. Philby was rewarded with a salary
of £1,000 a year from Socal, and took satisfaction in the Americans
being rewarded for their anti-imperialism.[3]

Thus Socal achieved their second, and far more valuable foot-
hold in the Middle East. They gained it paradoxically because of
their *lack* of previous involvement, which had avoided entangle-
ment in the Red Line Agreement, so that the richest prize of all
was left to the outsider.

The establishment of an all-American oil company in Saudi
Arabia was to change the whole Middle Eastern balance of power.
There was then no American diplomatic representation at all, and
there were only about fifty Westerners in the coastal capital, Jedda.
The arrival of the oilmen presented a new kind of Arabian wonder.
On the one hand was a desert kingdom ruled by an absolute
monarch with medieval autocracy, with about five million people,
many of them nomads. On the other hand were American tech-
nologists, as Philby described them, 'descending from the skies
on their flying carpets with strange devices for probing the bowels
of the earth in search of the liquid muck for which the world
clamours to keep its insatiable machines alive.'[4]

As the Bahrain oil came into production, and the first drilling
in Saudi Arabia showed promise, so Socal, from its isolated position, realised that it was very short of capital and marketing outlets,
many of which Exxon tightly controlled. Socal soon found a
partner in the only other of the seven sisters who was not bound
by the Red Line, Texaco, which was rapidly expanding under Cap
Rieber, had plenty of markets, including the new found customers
in Franco's Spain, and was glad of a new source of oil; and in 1936
Rieber bought a half share in both the Bahrain and Saudi Arabian

Concessions. Thus the two remotest companies, from the west and south-west of America, were now united in Arabia.

By May 1939 the Arabian oilfields were ready for production. The King made a journey across the desert to the new oil-town and his party camped outside in tents. There were two days of banquets and inspections, and the King and his party were entertained on the deck of the tanker *D. G. Schofield*, named after the founder of Socal. The King and his chief adviser were given automobiles, and congratulatory telegrams were read out from Rieber of Texaco and William Berg of Socal.[5] The King turned the valve on the pipeline, and the oil began to flow. He was so delighted that he soon increased the size of the concession to 444,000 square miles—as big as Texas, Louisiana, Oklahoma and New Mexico together.

It was a historic diplomatic occasion, but with no diplomats; the United States still had no representative, and all the negotiations had been between a country and a company. It was not until three months later that the U.S. Minister in Egypt, called Bert Fisher, was accredited to Saudi Arabia as well. The oilmen could well imagine that they were a private government, without the need of Washington. But four months after that royal ceremony, when the Second World War broke out, they soon felt themselves anxiously in need of diplomatic protection.

KUWAIT

At the same time another massive new source of oil had been discovered in one of the small independent sheikhdoms along the Persian Gulf that cut into the solid mass of Saudi Arabia, like bites in a corn-cob. Kuwait was then a small fishing and trading port, with few merchants owning fleets of dhows. The story of the Kuwait concession has taken a long time to emerge, revealing an even more devious battle between British and American interests, witht he governments supporting the companies behind the scenes.[6]

BP had taken a desultory interest in Kuwait since February 1914. The British government insisted in 1924 that BP must be given priority, but BP had plenty of oil from Iran, and they allowed their negotiating rights to lapse in March 1925. In the meantine Major Frank Holmes, fresh from his concessions in Saudi Arabia and Bahrain, had arrived in Kuwait and established a friendship with the Sheikh, who allowed him to look for oil. In 1927 he sold his Kuwait interests to Gulf, as part of the bargain package including Bahrain, for $50,000 (see p. 104). But after Gulf sold the Bahrain concession to Socal, they still retained their interest in Kuwait, which was outside the Red Line, and they retained Holmes to act for them.

BP remained unexcited by Kuwait: Violet Dickson, who arrived with her husband, the British Resident, in Kuwait in 1929, recalls that the company men in Iran treated the rest of the Gulf with some disdain. The Dicksons found signs of bitumen under the sand outside Kuwait, which was used by the Bedouins as a cure for camelmange,[7] but the BP geologists insisted as in Bahrain that the surface indications were wrong. BP nevertheless wanted to exclude Gulf, and the British government, as in Bahrain, was now insisting on a British nationality clause. But Gulf, on their side, were looking to Washington and the Open Door. After March 1932 Gulf's position was specially influential, for the Ambassador in London was none other than Andrew Mellon, who had helped set up the company, and whose family still owned a quarter of the shares.[8] After June 1932, when commercial oil was discovered in Bahrain, BP immediately became determined to get the Kuwait concession.

In Kuwait the representatives of BP and Gulf were now competing for the Sheikh's favours, providing a picturesque contrast: Holmes was acting for Gulf, a rugged and forceful man of sixty, who had become intimate with the Sheikh. The BP negotiator was Archibald Chisholm, a young man of thirty-two who, as Violet Dickson described him, was 'tall and languid, with a monocle and exquisite good manners'.[9] He had the British government behind him, but the Sheikh—having watched Bahrain and Saudi Arabia—wanted American participation.

While Holmes and Chisholm were tempting the Sheikh with rival offers, the two chairmen of the companies, Sir John Cadman and James Frank Drake, began discussing the possibility of acting jointly in Kuwait: neither wanted to bid up the price, or to upset the delicate balance of Middle East oil. By December 1933, after a year of talks, they agreed to negotiate jointly, and Holmes and Chisholm now returned to Kuwait as allies. But the Sheikh, as Chisholm observed, was now much enjoying the whole long haggle, and he suddenly revealed that he had been approached by another all-British company offering far better terms. The Sheikh was in fact secretly negotiating with a company called Traders, put together by a group of right-wing Conservatives, including

Lord Lloyd and Lord Glenconner, who wanted both to break into BP's territory, and to forestall American interests.

When Holmes learnt about it in October 1934, he invented an extraordinary story to dissuade the Sheikh: namely that Traders was a front for BP and Chisholm, devised to frustrate Gulf of its share.[10] Holmes thus persuaded the Sheikh quickly to work out a new compromise for the joint company, more beneficial to Holmes, keeping secret the fact that there was a genuine competitor for the concession. Thus deviously the joint deal was completed: on December 23 1934, Holmes and Chisholm signed with the Sheikh the historic agreement. BP's share in Kuwait had owed a great deal to the British government behind the scene, making nonsense of Cadman's contention that the company had succeeded commercially in spite of British government ownership.[11]

After two years of drilling in the wrong place, the joint company eventually in 1938 struck an oilfield which spurted out under its own pressure; and it has spurted out ever since. Frank Holmes, having initiated the transformation of three nations, shortly afterwards retired to a farm in Essex. He was called by the Kuwaitis Abu-el-Naft, the Father of Oil, and when he died in 1947 the Sheikh sent a huge wreath to his funeral. But he made no fortune out of the concession: the vast profits went to the companies which he had with such difficulty involved.

WAR AND CONTROL

The development of these two vast new sources of oil was soon set back by war. Two partners in Saudi Arabia, Socal and Texaco, became extremely worried about the vulnerability of their precious concession. King Ibn Saud was pressing them for money, particularly since his chief income, from pilgrims to Mecca, had completely dried up. The two tight-fisted companies faced the interesting problem of persuading governments to subsidise the King, while still keeping firm control over the concession. The resulting arguments between companies and governments brought up the old questions of the control of oil reserves, with echoes of Churchill's arguments over BP thirty years before—but with opposite results.

In the first years of the war it was the British who bore the main burden of subsidising the King. But the American companies became worried that the British might acquire too much influence with the King, particularly after the British had sent an expedition

to deal with locusts which, Texaco noted, included some mysterious geologists. They feared too that the old King, with his several
wives and twenty sons, might be succeeded by a less amenable
ruler.

In February 1943 the chairman of Texaco, 'Star' Rodgers (who
had succeeded Rieber) and the President of Socal, Harry Collier,
paid a momentous call on the Secretary of the Interior, Harold
Ickes who had two months earlier been given the new and powerful post of Petroleum Administrator for War. Ickes, 'the old
curmudgeon' as he liked to call himself, was very ambivalent about
oilmen, and in some ways typified the attitude of many Democrats
(then and later) who distrusted the whole business, yet felt compelled to co-operate. Ickes once wrote in his diary that 'an honest
and scrupulous man in the oil business is so rare as to rank as a
museum piece',[12] and he eventually resigned from Truman's
administration in protest against the appointment of an oilman,
Edwin Pauley, to be Under-Secretary of the Navy. But he tended
to be impressed by the oilmen that he actually met, and worked
closely with Teagle and others in the pre-war years.

The two oilmen, Rodgers and Collier, stressed to Ickes the value
of the Concession, and the dangerous influence of the British, who
had by now advanced around $20 million to the King. They asked
him to obtain lend-lease funds for Saudi Arabia. Their mission
was quickly successful. On February 18, President Roosevelt sent
a historic letter to Edward Stettinius, authorising lend-lease aid,
and declaring that 'I hereby find that the defense of Saudi Arabia
is vital to the defense of the United States'. More money began to
flow 'to which the Arabian response [wrote Philby] was a further
orgy of extravagance and misarrangement, accompanied by the
growth of corruption on a large scale in the highest quarters.' It
was a remarkable coup for the companies, but they soon found that
they had interested Ickes and the government rather more than
they had intended; in the words of Herbert Feis, the Economic
Adviser to the State Department, 'they had gone fishing for a cod,
and caught a whale.'[13]

For by the spring of 1943 Washington was becoming seriously
concerned—as they had been twenty-five years before—by the
prospect of a new extreme shortage of oil. The new thinking was
outlined by William Bullitt, the Under-Secretary of the Navy, in
a memo to Roosevelt in June 1943, explaining that 'to acquire
petroleum reserves outside our boundaries has become . . . a vital

interest of the United States'. Socal and Texaco, Bullitt reported, were worried that the British could persuade the King 'to diddle them out of the concession and the British into it', and they were therefore eager to get 'a direct American Government interest in their concession.' The fact that Iranian oil to the north was backed by the British government, provided a model for American government involvement in Saudi Arabia: 'We are forty years late in starting—but we are not yet too late.' Bullitt then recommended not (as the companies proposed) that the government should support the companies in return for a share in the oil reserves. Instead he advanced a much bolder move right out of the American tradition, and much influenced (as Bullitt made clear) by the example of BP: that the government should set up a 'Petroleum Reserve Corporation', to buy a *controlling* interest in Aramco, and to construct a refinery on the Persian Gulf. Bullitt stressed the point that would so often be stressed in the future: that oil should not be left to the oilmen. The new corporation would have no negotiators who had been employed by Socal, Texaco or Aramco. Their officials must be 'not only as above suspicion as Caesar's wife, but also appear to be'; and the government should drive 'as hard a bargain as possible with the present owners'.[14]

With remarkable haste, the idea of this new nationalised industry gathered weight. It was legally feasible through the existing machinery of the New Deal and Ickes wanted 'immediate and aggressive steps' to safeguard oil reserves. He insisted that the government must buy stock in the company, while the Secretary of State, Cordell Hull, was much less keen. But Ickes won the day, and by June 30 President Roosevelt had authorised the formation of a new Corporation, to acquire a hundred percent of the Arabian concessions. As Herbert Feis, the economic adviser to the State Department described it, 'the discussion with the President had been jovial, brief and far from thorough. A boyish note was in the President's talk and nod, as usual when it had to do with the Middle East.'

The new Corporation had Ickes himself as President, the Secretaries of State, War and the Navy among the directors, and Abe Fortas as Secretary. It had its first meeting on August 9: among others present was John J. McCloy, the shrewd lawyer-administrator who was Acting Secretary of War. Few corporations have ever had such powerful directors; but they soon ran into difficulties. Socal and Texaco, not surprisingly, indignantly rejected the proposal for outright ownership and would only consider selling a one-third interest.

Ickes and his board were now eyeing the vast reserves of Saudi Arabia with growing excitement and they sent out an expedition to the Persian Gulf, headed by Everett de Golyer, the eminent oil geologist; and de Golyer's report, when he returned, confirmed their involvement. 'The centre of gravity of the world of oil production,' he predicted, 'is shifting from the Gulf-Caribbean areas to the Middle East, to the Persian Gulf area, and is likely to continue to shift until it is firmly established in that area'. The full significance of this bold and accurate prediction was to take many years to sink in.

Ickes still thought he could make a deal with the oil companies. In early October he met with Rodgers of Texaco, the toughest partner, who promised a satisfactory agreement; but a week later Texaco withdrew all proposals, and repudiated all arrangements. What exactly happened between Ickes and Rodgers has been much disputed. Rodgers maintained that it was Ickes who abruptly called it off; Ickes maintained that Texaco were 'pollyfoxing' him while lobbying behind the scenes, and that he decided to call Rodgers' bluff. Ickes had little doubt of the real reason for Texaco's change: when the first negotiations had begun, it looked as if the Germans might over-run the Middle East. But by mid-1943 Rommel had been ejected from North Africa, and Texaco no longer needed the government's support.

Could Ickes' plan have ever worked, given more of a chance? Opinions today vary widely: to George McGhee, the Texan oilman who became an Assistant Secretary at the State Department after the war, the idea was a non-starter. 'It's not the American way,' he told me, 'only Ickes was really for it'; but Abe Fortas still thinks it should have been pressed through: 'at least it would have given the government a seat at the poker-table'. 'It might have been a better solution in the future: at least it couldn't have worked out much worse.'

THE CRITICAL PIPELINE

Ickes still did not give up his plans for control. He made a speech to the oilmen pointing out that most other nations had government-controlled oil companies, and decided on another daring expedient. The government should construct a thousand-mile pipeline, to carry the Saudi Arabian oil to the Mediterranean;

and in return the companies would guarantee 20 percent of the oilfields as a naval reserve, which would be available to the Navy at cut price. This plan had huge diplomatic implications for, as Karl Twitchell observed, 'it committed our government to a fixed foreign policy for at least twenty-five years.[15] But the war leaders were anxious to reinforce the American presence, and they heavily backed it ('never did a pipeline', wrote Feis afterwards, 'have so distinguished a diplomatic and military guard of honour'). It certainly suited Texaco and Socal, who could now carry their oil much more cheaply, and Rodgers and Collier, the two Presidents, duly signed an agreement with Ickes.

This *fait accompli* provoked uproar; partly from the British, who saw a threat to their Middle Eastern dominance; and more dangerously from the rest of the American oil industry, who saw their rivals gaining an unfair advantage. The oil industry was mobilising itself against Ickes with a special committee to formulate national oil policy, which proclaimed that governmental interference was harmful 'to this most individualistic of all economic activities, the oil industry'. Senator Moore of Oklahoma described the government pipeline as a venture in imperialism, and a report sponsored by the Industry's War Council described the proposal as revealing 'the fascist approach'. Eugene Holman of Exxon (which was in fact to join the Aramco group), insisted that the United States was more than self-sufficient in oil, with enough reserves to supply needs at the present rate for a thousand years. Texaco and Socal in the meantime wisely laid low, leaving the government to fight for them.

The pipeline was fraught with diplomatic implications and set off a whole new argument across the Atlantic. Ickes and Roosevelt tried to interest Churchill in an agreement to recognise American interests in the Middle East. Churchill cabled Roosevelt that 'some British quarters' suspected America of wanting to deprive Britain of her oil interests, and Roosevelt replied that there were rumours that the British were trying to horn in on Saudi Arabian oil. Eventually the British appointed a cabinet delegation headed by Lord Beaverbrook, which met with Cordell Hull and Ickes and actually signed an agreement on August 8, 1944. It was carefully vague, but enough to raise a new storm among American oil-producers against this 'super-cartel'. It was then rejected by the Senate, redrafted and renegotiated by the British Labour government, attacked by the American oilmen, and finally defeated again in the Senate in July 1947. Any hope of an agreed Anglo-American policy collapsed. In view of the gathering rivalries, that was scarcely surprising.

Ickes' government pipeline was in the end sabotaged by its enemies. Texaco and Socal decided to build it themselves, and in July 1945 they organised the Trans-Arabian Pipeline Company, known ever since as Tapline. The planning of the route involved still more difficult diplomacy with the mounting tension between Arabs and Jews. The construction was interrupted by the first Arab–Israeli war: it was not till 1949 that Syria and Lebanon agreed to let the pipeline be built. The other American sisters were furious that the Tapline was being allowed steel which they desperately wanted, but it at last went ahead, a steel snake thirty-inches wide, laid through the desert, costing $200 million. By November 1949 the first oil was being pumped into the terminus at Sidon in Lebanon to be loaded onto tankers for Europe.

The pipeline was to have a chequered history; it provided an ideal target for guerrillas, an instrument for boycott, and a bargaining counter for Syria against America: passing through three countries, it could never really be safe, and in April 1975 it was closed down. But its diplomatic significance for Texaco and Socal was enormous. It was not just a means of carrying oil cheaply to Europe; it involved a commitment of U.S. foreign policy for over a quarter of a century, as Twitchell predicted—without any governmental control. The effect of Ickes' original campaign had been to help the companies when they most needed it, and then to leave the running of the concession, with its vast political implications, in the hands of the companies.

THE CLOSED DOOR

By the end of the war the dominant influence in Saudi Arabia was unquestionably the United States. King Ibn Saud was regarded no longer as a wild desert warrior, but as a key piece in the power-game, to be wooed by the West. Roosevelt, on his way back from Yalta in February 1945, entertained the King on the cruiser *Quincy*, together with his entourage of fifty, including two sons, a prime minister, an astrologer and flocks of sheep for slaughter. The King brought gifts of swords and daggers, and the President promised him an aeroplane. Three days later, Churchill entertained the King at the Hotel du Lac, at Fayoum Oasis; he, too, was given jewelled swords, and hastily promised an armour-plated Rolls-Royce. But Churchill's reference to the King's 'un-failing loyalty' could not conceal that he now looked across the Atlantic both for his defence and his income from oil.

There was already a cloud looming over the new-found American relationship, for the question of the Jews had inevitably come up in the talk between Roosevelt and the King. In a sub-sequent letter to the King, Roosevelt reaffirmed that the United States would not change its policy towards Palestine without consulting the Arabs. But two months later, Roosevelt was dead, and President Truman was soon giving full support to the estab-lishment of the new State of Israel: he agreed to make public Roosevelt's letter to the King, but would not reaffirm his predeces-sor's statement of neutrality and friendship.

There were quickly indications of trouble from Saudi Arabia; and in December 1946 the King's son, Amir Feisal (later King until March 1975) came to Washington to see Truman to make clear his father's worries about Zionism. Dean Acheson had a foretaste of a personality that was to loom large in American foreign policy:

> The Amir, striking in white burnoose and golden circlet, which heightened his swarthy complexion, with black, pointed beard and mustache topped by a thin hooked nose and piercing dark eyes, gave a sinister impression, relieved from time to time by a shy smile . . . As he talked with President Truman, it seemed to me that their minds crossed but did not meet. The Amir was con-cerned with conditions in the Near East, the President with the conditions of the displaced Jews in Europe . . . The Amir im-pressed me as a man who could be an implacable enemy and who should be taken very seriously.[16]

Two opposite American foreign policies were thus both firmly recognised; support for the State of Israel, which was critical for honour and votes, and support for Saudi Arabia, which was critical for oil. The State Department's solution, as we will see, was to delegate their diplomacy in the oil countries as far as possible to the companies, and to regard them as an autonomous kind of government: and through this means the two policies were kept remarkably separate for the next twenty-five years.

Socal and Texaco were very aware of their diplomatic indepen-dence, and of the fact that they were sitting on the biggest and cheapest source of oil in the world. De Golyer and other geologists upped their earlier estimates until it became clear that the Saudi desert had more oil underneath it than the whole of the United States. Once the wellheads and pipelines had been installed, the oil, under its own pressure, pushed its way to the sea, to fill the waiting tankers at the new port of Ras Tanura, or through the thousand-mile pipeline to the Mediterranean.

Socal and Texaco, through good luck and the timidity of the others, had stolen a march on all the other five sisters. This was very clear to Exxon's new president, Eugene Holman, who took over in April 1944; for Holman, a geologist with long experience in Latin America, liked to pride himself on his global statesmanship. He realised that Exxon's supplies from Iraq, which had first confined them inside the Red Line, were puny compared to the new bonanza. As Europe and Japan recovered from the war, and began to need far more oil, so Exxon and its little sister Mobil became fearful that Socal and Texaco would undercut them in the world markets with cheap Arabian oil.

Here, it might seem, was a real opportunity for the outsider to break the old dominance of Exxon and to inaugurate a new era of genuine competition in the Middle East. But that was not how the oil industry worked, and Socal and Texaco, on their side, had worries in the midst of their new wealth. They needed more capital, to reassure the King that the oilfields were being adequately exploited; and they needed access to more markets than they themselves could supply. Moreover, the two companies still felt very unsure of the political future in this remote part of the world: they dreaded a Communist take-over, or an unfriendly successor to Ibn Saud, and they felt the need for diplomatic support.

For these reasons Socal and Texaco considered bringing two more American sisters into their Arabian partnership and the way was prepared for a new carve up. Not all the directors agreed; Ronald C. Stoner, a director of Socal, argued in June 1946 that the two should continue to go it alone, making contracts with other oil companies to provide outlets, and challenging the markets of Exxon.[17] But he lost the argument: Socal commissioned a study which found that they would make more money out of a smaller share of a bigger venture. The sisters once again felt safer in each other's company; and by September 1946 Harry Klein, the new President of Texaco, was already discussing with Exxon and Mobil the possibility of joining Aramco.

Mobil, on their side, desperately needed crude oil to fill their markets. But they had doubts about the potential of the Saudi

Arabian reserves, and were also worried about the political implications of the new collaboration. Their chief counsel George
Holton sent a candid memo to their President, Brewster Jennings,
in October 1946 warning him about the anti-trust aspects of an
arrangement which 'would place practical control of crude
reserves in the hands of seven companies'; and he went on to
reflect: 'I cannot believe that a comparatively few companies for
any great length of time are going to be permitted to control world
oil resources without some sort of regulation.'[18] His scepticism
was premature, but prescient.

There still remained the problem of the old Red Line Agreement, which after twenty years still restricted both Exxon and
Mobil. They had both tried before the war to escape from the
rule, but Gulbenkian and the French would not let them. Now,
supported by the State Department, they tried to break the agreement, referring to the Open Door. Their representatives, Orville
Harden and Harold Sheets, came to London to negotiate with the
British partners, and there followed a bizarre new round of horse-
trading between the sisters. BP was worried that the Arabian oil
would undercut them; but Exxon could appease them by promising a long-term contract to buy their oil from Kuwait and Iran,
and to collaborate in building a new pipeline to the Mediterranean.
Shell's similar worries were mollified by promises of contracts from
Mobil.

But there was still the French company CFP and the obstinate
Gulbenkian, whose famous 5 percent was now yielding him around
$20 million a year. Exxon was confident that their claims were now
invalid, since their shares had been declared forfeit during the war,
when both the French company and Gulbenkian were under
enemy occupation. But Gulbenkian, from the Ritz in Paris, was
determined to fight it out, with the help of his eminent lawyer, Sir
Cyril Radcliffe. The French joined him in what promised to be the
bitterest and most lucrative law suit in the whole litigious history
of oil.

The American companies, dreading the hostile publicity, soon
looked for a compromise. They enlisted the help of the American
government, which in turn pressed the British: 'The damage to
international amity likely to result from a collapse of the present
negotiations [they warned the Foreign Office] greatly exceeds any
conceivable detriment to private interest . . .' There was some
worry within the State Department about the magnitude of the

proposed new carve-up between the seven: and Paul Nitze proposed a more competitive solution. Mobil should buy Exxon's
12 percent share in Iraq Petroleum, while withdrawing from
Aramco. This, he pointed out, would retard the growing consolidation of the two biggest American oil companies, and would
help to reassure the smaller American companies who were
alarmed that Middle East Oil was 'being pre-empted by various
combinations of large American and British companies'.[19] But the
sisters would not consider this, and in the meantime the massive
court case was building up.

What the French really wanted was a share in the Saudi Arabian
concession, which they knew was far bigger than Iraq's. The
American companies eventually persuaded them to accept a compromise, by which Exxon would help with a massive expansion in
Iraq, building two pipelines to the Mediterranean and increasing
production sixfold. Gulbenkian still resisted, even though his income would be massively increased by the Iraq expansion, and
eventually he had to be paid off with a further allocation of free oil,
until he finally agreed to cancel the Red Line Agreement.

Finally in November 1948, the day before the great lawsuit was
due to begin in London, representatives of the chief companies—
BP, Shell, CFP, Exxon and Mobil—assembled in the Aviz Hotel
in Lisbon, where Gulbenkian was now living; and the agreements
were finally signed at two o'clock in the morning.[20] Thus the Red
Line was finally erased, and Exxon and Mobil were free to enter
Saudi Arabia. Ostensibly it was another victory for the Open Door;
but once again, as soon as the Americans were inside the door
slammed behind them—this time behind the most protected and
richest concession of all—leaving other countries to nurse their
resentments for years to come: 'The French', concluded a Senate
Report twenty-five years later, 'never forgave the Americans for
keeping them out of Saudi Arabia.'[21]

The price of entry seemed at the time very high. Exxon and
Mobil were first offered 20 percent each, but Mobil, after bitter
arguments on the board, decided that the cost was too high for the
dubious benefits, and took only 10 percent—a misjudgment so
huge that it has been the subject of recriminations ever since.
Exxon therefore gladly took 30 percent, becoming co-equal with
Socal and Texaco, and acquiring what it had sought: a position of
strength in the new 'centre of gravity' of oil. It had got it, not
through any great enterprise, but once again through its control

of the markets. The total cost to Exxon and Mobil—payable partly in revenues from the oil—was later reckoned at half-a-billion dollars. But the price was trifling in view of the return. The Saudi Arabian oil proved to be the biggest bargain in the history of the business.

In the meantime, there rose up in the desert, on the edge of the oilfields, the most amazing of all company towns: the compound at Dhahran which provided the headquarters of the Arabian American Oil Company, or Aramco. Bungalow houses sprang up in neat rows, with creepers up the walls and green lawns alongside the desert, and a complete suburb formed itself with a baseball park, a cinema, swimming-pools and tennis-courts. It was an astonishing optical illusion, looking like a small town from Texas or California, whence many of the inhabitants came; except that it was ringed round with a high barbed-wire fence, with beyond it an expanse of limitless desert, with only a few oilwells and pipelines to break the monotony.

In this isolated outpost the employees, or 'Aramcons' as they called themselves, soon acquired a special kind of character or nationality of their own, caught between loyalties to two countries, and to four companies: as one of their early employees described it: 'Aramco was, in effect, the neurotic child of four parents, subject to the whims, qualms and jealousies of each.'[22] The sense of uncertainty in this desert community was increased by the anger of the Arabs at American policy towards Israel, and the eccentric demands of the Saudi Kings. The Aramcons took pride in the fact that they were pragmatists, simply providing technology and oil for the good of the world, and they were proud, too, of their growing contributions to Saudi Arabia's social services, building hospitals, schools, roads and industries for the nation. They were determined to avoid the paternalistic and imperial attitudes of BP up in Iran, and they made fun of the British officials in Bahrain, the 'lion-tamers' in their topees.[23] But however much they tried to avoid politics, they were in one of the most politically exposed positions in the world—a Western enclave in the midst of an autocratic kingdom. They were champions of the Arabs, from a country committed to support for Israel. And they were operating one of the most special of all special relationships, between a company and a country.

THE PRICE

Just when the Middle East was opening up its cheap oil, the price of gasoline in the United States was shooting up, with the ending of wartime controls. The price of oil suddenly again became an explosive political issue, and there was a new surge of populist resentment against the old bogey, the oil cartel.

The four partners in Aramco were in some disagreement as to how they should price their cheap Arabian oil. Exxon wanted to put it up far enough to be in line with its other oil suppliers, and to keep it out of the U.S. market; but there were objections from the original partners. A memo from Socal and Texaco in June 1947 warned that the Europeans (to whom much of the oil would be sold) were resentful of the high price of oil: 'The people of Europe in general have had a despairing attitude towards their future rehabilitation, if not survival . . . An increase in prices of Middle East crude now would produce, in our opinion, a widespread reaction that we are squeezing and taking advantage of them in their desperate straits.'[24] Within Exxon itself, the planning staff wrote a warning to President Holman: 'In the Middle East, where the oil companies are large in size and few in number, and where partnership arrangements are very common, it is particularly important to avoid anything which would look like monopolistic abuse . . . Foregoing localised high returns . . . would be an investment in future sound and satisfactory relationships.' But the board of Exxon were determined to maintain a high price, and after bitter argument and the threat of legal action, the other partners eventually agreed to put up the price from $1·02 to $1·43.[25]

Already before Exxon joined Aramco, there had been a major row about the oil price. Early in 1947 it transpired that Aramco was contracted to sell oil to France at 95 cents a barrel, and to Uruguay at $1 a barrel, while it sold oil to the U.S. Navy for up to $1·23 a barrel. A Senate Committee, headed by Senator Brewster of Maine, then held hearings which brought into the open much of the devious wartime dealings of Texaco and Socal. The timing was poignant for Exxon and Mobil were just contracting to acquire their shares in Aramco, and their contract was agreed just two days before President Truman announced the 'Truman doctrine', for support of Turkey and Greece, which made Middle East investments appear substantially safer.

The oil companies thus appeared to be getting rich on the backs of the government, and in April 1948 the Brewster Committee published a devastating attack on Socal and Texaco: 'The oil companies have shown a singular lack of good faith, an avaricious desire for enormous profits, while at the same time they constantly sought the cloak of United States protection and financial assistance to preserve their vast concessions.' The committee noted the number of government oil experts who had formerly been employed by oil companies, and recommended that Congress should establish a special petroleum board and continue its investigation: the Attorney-General, they said, should give the utmost consideration to the question of Exxon and Mobil joining Aramco, 'with its possible effect of lessening competition'. The companies, in a memorandum, replied that they had contributed and risked far more in Arabia than the U.S. government, to which they had no special obligation. The Aramco sisters could now afford to flaunt their independence.

There was soon a more resounding outcry about prices, as a result of the launching in 1948 of the Marshall Plan and its instrument, the European Co-operation Administration (ECA). Paul Hoffman, the administrator, discovered that American oil companies were charging the ECA much more for their oil than they were charging their own affiliates; and Eugene Holman of Exxon when questioned, had to explain how the pricing of oil was still based on the 'Gulf-plus' system that had been consecrated at Achnacarry Castle twenty years before, and on the eerie concept of 'phantom freight'. As he put it in inimitable oil jargon:

> our announced FOB prices for crude oil supplies at the Eastern Mediterranean or Persian Gulf are equivalent to the Caribbean price for crude plus freight at published United States Maritime Commission rates from the Caribbean to Western Europe less freight on the same basis from either the Eastern Mediterranean or the Persian Gulf, depending on the supply point to Western Europe.

In other words, the cheap Arabian oil was fixed at the high American price.

The oil companies eventually agreed under pressure to establish an alternative 'basing point' for oil prices at the Persian Gulf. Oil from the Middle East would therefore not be charged with phantom freight. provided it did not travel further than the mid-Mediterranean; but the oil from the Persian Gulf was still to be charged at the same rate as the more expensive American oil. This formula in turn was modified as the Middle East oil increased in volume; the 'equalisation point' was moved from the mid-Mediterranean first to London and then to New York.

In early 1949 Hoffman charged the four members of Aramco with differential pricing, and instituted his own investigation. As a result the oil companies soon made cuts in their prices. But the question of oil prices continued to reverberate through Washington, with a succession of investigations. The old mole of anti-trust, having gone underground during the war, was coming up again to the surface.

Even with the cuts in prices, the sisters were making vast profits from Middle East oil. 'It is difficult to say what would have happened,' commented one oil economist, Paul Frankel, at the time, 'had production in the Middle East not been in the hands of companies which had strong interests in the Western Hemisphere.'[26] But for the time being, the sisters were safe: the new oilfields had been divided between the existing companies, nearly all of whom (except BP) also produced oil in America, so that they would not let the new oil undercut the Americans. Their interests were interlocked, in Iraq, Kuwait, Bahrain and Saudi Arabia. So that though their competition in marketing was fierce, they all shared the same concern for maintaining high prices and avoiding an uncontrollable glut: between them they could avoid opening up too many new fields.

And meanwhile the world was clamouring for the stuff, and the car-factories and highways were opening up Europe to a great new age of oil, transforming the landscape and way of life. In spite of the high profits, the oil was the cheapest of all fuels, soon threatening the coal-miners with redundancy, and providing a cheaper, cleaner and far less arduous alternative.

With America booming and Europe reviving, and with oil profits constantly under fire, few people were anxious to suggest that oil was not too expensive, but too cheap; or that the world was relying on a fuel whose supplies would be increasingly uncertain. The old optimism of the Pennsylvania drillers still seemed justified; just when a shortage seemed imminent, a vast new oilfield would open up. If one foreign country, like Mexico became awkward, there was always another that was glad to make a deal, like Saudi Arabia or Venezuela. From their position of power, the oilmen were slow to realise that the producing countries, too, were gradually becoming closer to each other, and discovering their own potential strength.

FIFTY-FIFTY

It was on the other side of the world that nationalism first seriously confronted the oil companies, and it happened in the middle of the Second World War, when it was hard for them to withstand it. The Mexican crisis had apparently been overcome, with the help of Venezuelan oil. During the war years Venezuela, with an area bigger than Texas and a population of only six million, had rapidly become the chief exporter of oil in the world, and a vital source for the three companies most involved, Exxon, Shell and Gulf. The war in Europe—though the British public were unaware of it—was dependent on Venezuelan oil.

Venezuela had been transformed through oil into the richest country in Latin America: the capital, Caracas, filled up with automobiles and between 1920 and 1936 its population doubled. But the position of the companies had been basically unstable, as in Mexico, particularly since the collapse of the dictator Gomez in 1936, and they were attacked for their exorbitant profits and the wretched conditions in the oil slums. By 1938 Venezuelans, under the military regime of General Medina, demanded a revision of oil contracts, to allow higher royalties and taxes, in return for a forty-year renewal. They threatened the companies with national-isation if they would not agree. The companies were at odds: Shell, under its diplomatic manager, John Loudon, saw advantages in coming to terms; but Exxon had a major boardroom row, with the old guard encouraged by their local manager in Venezuela, appalled at the threat to 'the sanctity of contracts'. But the State Department were in no mood to back up Exxon's militancy, and were conscious of Europe's desperate war-time need for the oil. In contrast to Mexico, the way was open for a peaceful compromise. They recommended two mediators, including Herbert Hoover (the son of the ex-President) to help draft a new law. In fact the oil companies were soon glad to pay higher royalties in return for stability and a forty-year contract. They still made ample profits. They stepped up exploration and development, and soon doubled their output of oil.

In 1945, the radical party Acción Democrática came to power; and with it a new kind of oil minister, a scholarly and monkish economist called Perez Alfonso, who really understood the economics of oil. He had watched closely the developments inside the United States, the drastic depletion of reserves, the greed of the companies, and the enforcement of conservation and rationing in Texas. He saw Venezuela's problems in the larger perspective of world supplies in future decades, and he was to become the chief architect of OPEC.

Alfonso insisted that the Venezuelan government should have a fifty-fifty share in all oil profits. A basic new law was passed in November 1948 — just before a new coup removed Alfonso temporarily from office. But the new law remained in force and the companies soon realised that it gave them greater security. It established the government as their partners, and it made foreign-owned corporations much less vulnerable to nationalist attacks. The fifty-fifty arrangement soon became a rallying cry for other oil-producing countries, and the idea quickly crossed the six thousand miles to the Persian Gulf.

THE KING'S TAX

In Saudi Arabia, it was not long before the old King was demanding a bigger share in the profits. He was becoming increasingly extravagant: his huge family began travelling to America, bringing home large cars and gadgets, and gourmet foods were flown in by plane to Riyadh. Aramco were obedient to his whims, and built hospitals, schools and roads to try to satisfy him. He wanted a railroad from the capital to Dhahran, the oil town, which was reckoned to be quite uneconomic, but Aramco provided it, at the cost of $160 million, advanced out of future royalties. But the King still clamoured for more revenue and asked why he, too, could not receive a 50 percent tax as in Venezuela. Relations between the King and Aramco became strained, particularly after two newcomers had bought concessions on the edge of his kingdom (see Chapter 7) on far better terms. Aramco were fearful of losing their concession, but reluctant to give up any of their huge profits.

Aramco's officials met with the State Department in November 1950. They had a sympathetic hearing from the Assistant Secretary George McGhee, the Texas oilman and son-in-law of the geologist Everett de Golyer; McGhee was convinced of the need to give the Saudis more income,[27] and the State Department was worried by the Communist danger in the Middle East.

The State Department and Aramco then agreed on a scheme of

beautiful simplicity. Additional payments to the King, should be regarded henceforth as constituting a foreign income tax, so that under the existing rules for double taxation, they would not be taxed inside the United States. The King's share would simply be deducted from the company's tax bill. There were misgivings for some State Department officials about 'what in effect would amount to a subsidy of Aramco's position in Saudi Arabia by U.S. taxpayers'.[28] The 'Golden gimmick', as it was later called, deprived the United States Treasury of $50 million in taxes the very next year, enriching the King by the same amount: and as Saudi oil production soared, the loss of taxes became much more spectacular. But the tax device suited the State Department as well as the King and Aramco, for it was really a means to provide foreign aid to a kingdom which was important strategically, without having to submit it to Congress: a particularly embarrassing procedure when Israel was struggling for survival. It was not until hearings six years later that the tax device was publicly aired.

Other oil companies and countries were soon forced to adopt the same tax dodge to compete with Aramco. This technical change was to have vast diplomatic consequences, and helped to change the basic balance of the big companies. It provided a huge inducement to invest abroad, rather than inside America, to minimise taxation; so that by 1973 the five American sisters were making two-thirds of their profits abroad, and paying no taxes to the United States government on those earnings.[29]

It also changed the internal accounting of the companies, who soon realised that the more profit they made from their business 'upstream', producing oil abroad, and the less profit they made 'downstream', refining and selling oil to the consumers, the less tax they would pay in the United States. They thus took little trouble to make money out of filling stations, which they regarded primarily as outlets for their flood of oil. They littered the freeways with them, to attract customers at all costs, while the dynamic energy of the companies went into production and exploration. Engineers and geologists dominated the boards, while marketing men were at a discount: the companies were preoccupied with the simple word, Crude.

The new fifty-fifty arrangement had another result with far-reaching consequences which few oilmen foresaw. Because the producing countries were now partners in the profits, they soon insisted that the crude oil must be sold at a price which must be publicly fixed, and not buried in the companies' accounting. Accordingly, the companies agreed to publish a 'posted price' at which they would offer their oil for sale to anyone; and on that price would be based the taxes paid to each government. It seemed at the time a fair system, but it had a serious drawback. The countries became accustomed to a steady income derived from the fixed price, and could not imagine it coming down, so the posted price soon became an artificially high price on which companies paid their taxes.

The producing countries thus were increasingly insulated from the real market, and the companies went on paying taxes on the basis of the posted price, not the real market price, and obtaining tax relief on that basis. In terms of the tax laws, this was flagrantly improper. The producer countries were now really receiving taxes based not on profits, but on sales; and this relief allowed companies to pay lower U.S. taxes than any group of industries. In 1972, for instance, Exxon paid out of its global income only 6·5 percent in U.S. taxes, and Mobil only 1·3 percent. In the mid-sixties the Internal Revenue Service tried to question the pricing system, which was costing huge sums to the Treasury. But the attorney representing the five American sisters, John J. McCloy (who will emerge later in Chapter 8) reminded the then Secretary of State, Dean Rusk, of the real rationale of the tax allowance: 'If the companies did not provide the necessary revenues by paying substantial taxes to producing countries, large amounts of direct foreign aid might well be required.'[30]

Thus the role of the oil companies in foreign policy was firmly underlined: they were given private privileges to enable them to be the paymasters of the Arab states. It was from the State Department's point of view a neat, even brilliant solution for they could overtly support Israel, and covertly support the Arabs, effectively bypassing Congress. But it was a solution which also served greatly to increase the power of the oil companies, to which the government had virtually delegated part of its foreign policy. It was a power which they were not slow to exploit.

NOTES

1. Henry St. John Philby: *Arabian Jubilee*, London, 1962.
2. Karl Twitchell: *Saudi Arabia* (3rd edition), Princeton, 1958, p. 221.
3. Elizabeth Monroe: *Philby of Arabia*, London, 1973, p. 205.
4. Philby: p. 179.

5. Wallace Stegner: *Discovery!* (Export Book), Beirut, 1971, p. 108.

6. For this section I am much indebted to Archibald Chisholm, both for his private recollections, and for access to his forthcoming authoritative study of the period: *The First Kuwait Oil Concession Agreement*, Frank Cass, London, 1975.

7. Violet Dickson: interview with author in Kuwait, February 1975.

8. Because of Mellon's interest, negotiations were conducted by his Minister, Ray Atherton, but Mellon's influence was evident in the background, and at one point he personally sent a stiff note to the Foreign Office for which he was reproved by the State Department. Chisholm, pp. 134–5.

9. Violet Dickson: *Forty Years in Kuwait*, London, 1971, p. 117.

10. See Holmes' cable to Gulf in London, October 15, 1934 (Chisholm, p. 208).

11. See page 88.

12. Harold L. Ickes: *The Secret Diary of Harold L. Ickes, 1933–1936*, New York, 1953, p. 646.

13. Herbert Feis: *Seen from EA*, New York, 1947.

14. Senate Multinational Subcommittee: *A Documentary History of the Petroleum Reserves Corporation, 1943–44*, Washington, 1974, pp. 4–5.

15. Twitchell: p. 237.

16. Dean Acheson: *Present at the Creation*, Signet, New York, 1969, p. 241.

17. Multinational Report: *Oil Corporations and U.S. Foreign Policy*, Washington, 1975, p. 47–8.

18. George Holton to Brewster Jennings, October 28, 1946, Multinational Hearings: 1975, part 8.

19. Multinational Hearings: Part 8, February 21, 1974.

20. The meeting is vividly described by Nubar Gulbenkian: *Pantaraxia*, London, 1965, p. 227.

21. Multinational Report: 1975, p. 55.

22. Michael Sheldon Cheney: *Big Oilman from Arabia*, London, 1958, p. 149.

23. Stegner: p. 47.

24. Multinational Hearings: Part 8.

25. Multinational Report: 1975, p. 80.

26. FTC Report: 1952, p. 355.

27. Multinational Hearings: Vol. 4, p. 89.

28. Ibid: Vol. 8.

29. Multinational Hearings: Vol. 4, pp. 12, 95.

30. Multinational Report: 1975, p. 93.

6

Iran and Democracy

Never had so few lost so much so stupidly and so fast.
Dean Acheson, 1974[1]

This last British empire, the empire of oil, has 'paid' better than any other.
John Strachey, 1959[2]

Sir William Fraser, who took over from Cadman as chairman of BP in 1941, was a man with few doubts about the special national role of his company. He was a silent, craggy Scotsman, with an intimidating Glasgow accent and a bleak sense of humour. Men in government were inclined to accept the fact that he knew more about oil than anyone else. He had been born into oil: he inherited from his father the biggest company in the then-prosperous Scottish oil-shale industry, and later merged it with six other companies into BP, to provide them with Scottish outlets. He thus moved into BP fortified with a block of shares and soon joined the board. He first became a key figure as the money-man for Sir John Cadman; helping to negotiate the new agreement in Iran in 1933, but he lacked Cadman's breadth of outlook. After the war Sir William was slow to recognise the changing shape of the world around him, particularly after Britain had given independence to India in 1947—a fact which fundamentally changed the military position in Iran next door. Sir William was convinced that BP could and should hold on to its monopoly of Iranian oil, defying the growing nationalist temper, and his intransigence led to the biggest political upheaval in the history of oil, which was further to alter the balance of power in the Middle East. It also sharply raised the questions that were inherent in the character of BP, with its half-government ownership. Who was really in charge of oil policy? And was the government prepared to use force to protect its interest?

It began as an exclusively British dilemma. BP had emerged after the war with a far greater importance to Britain, as the

cornerstone of Middle East oil. In spite of its remarkable failure to exploit Bahrain and Saudi Arabia, BP had spread out from its original territories in Iran and Iraq down the Persian Gulf to Kuwait, and it was reaping still greater profits for its shareholders, notably the British government. Ever since 1923 it had needed to raise no more outside capital, ploughing back its own profits to build up a world-wide market and fleet of tankers, and after the Second World War it regularly declared dividends of thirty percent.

BP was now in a much more promising position than its old rival, Shell, which had no oil in the Middle East except for its share in the Iraq consortium. Shell's oilfields in Mexico were nationalised, while its fields in Venezuela which had been so crucial in the war, now had a limited life compared to BP's vast reserves in Iran and Kuwait. The departure of Deterding, however welcome, had left Shell to be run by his minions and yes-men, with a cautious committee at the top. It did not begin to recover its confidence until the arrival of a tough Welshman, Sir George Leigh-Jones, who dominated the company in the post-war years.

But BP, with its Middle East strongholds, had acquired a confident regimental character of its own, happily combining a sense of public service with the fruits of huge profits, and untroubled by the political outbursts and anti-trust crusades that embarrassed the American sisters. Sir William referred to civil servants contemptuously as 'The gentlemen in Whitehall'. Sir Charles Greenway became Lord Greenway, Sir John Cadman became Lord Cadman. The British government were glad to accept the profits, and were told as little as possible. The two government directorships were regarded as sinecures for retired public servants; in 1951 they were held by Field Marshal Lord Alanbrooke, and Sir Thomas Gardiner, a former head of the Post Office.

Iran remained the jewel in the crown of BP, and its supplies seemed limitless. During the war, after a period of uncertainty, production was pushed up to help fuel the allied armies; and in the first five years after the war production nearly doubled. As the old Empire disappeared, so the commercial empire of BP seemed all the more remarkable and necessary, bringing benefits to both British and Iranians. The company had created, in the words of the official biographer, 'not only a flow of oil but a way of life',[3] and the two thousand British employees saw themselves as bring-

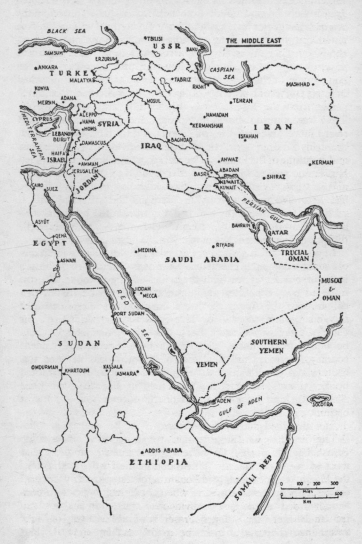

THE MIDDLE EAST

BLACK SEA
SAMSUM
ERZURUM
ANKARA
TURKEY
KONYA
MERSIN   ADANA
MALATYA
CYPRUS
MEDITERRANEAN SEA
LEBANON
BEIRUT
HAIFA
ISRAEL
JERUSALEM
AMMAN
JORDAN
CAIRO
SUEZ
ASYŪT
QENA
EGYPT
ASWAN

TBILISI   U S S R   BAKU
CASPIAN SEA
TABRIZ
RASHT
TEHRAN
MASHHAD
MOSUL
ALEPPO
HAMA
HOMS   SYRIA
DAMASCUS
HAMADAN
KERMANSHAH
ISFAHAN
I R A N
IRAQ
BAGHDAD
AHWAZ
KERMAN
BASRA   ABADAN
KUWAIT
KUWAIT
SHIRAZ
PERSIAN GULF
BAHRAIN
QATAR
TRUCIAL OMAN
MUSCAT & OMAN

MEDINA   RIYADH
SAUDI ARABIA
JIDDAH
MECCA
PORT SUDAN
RED SEA

S U D A N
OMDURMAN   KHARTOUM
KASSALA
ASMARA

SOUTHERN YEMEN
YEMEN
ADEN   GULF OF ADEN   SOCOTRA

ADDIS ABABA
ETHIOPIA
SOMALI REP

0   100   200   300
Miles
500
Km

ing employment, houses, schools, hospitals to Iran; why should Iranians not be grateful? Yet, in the disaster that followed, in the words of Brigadier Longrigg, a former IPC official who later wrote a standard work on oil in the Middle East, 'the whole of this effort, the whole of a half-century of generous and enlightened treatment of its own workers and the public was, in the final destiny of the company in Persia, not only treated as of no account, but attacked in terms suggesting not mere neglect but the crudest exploitation'.[4]

But from the Iranian side, this picture was unrecognisable. They were a proud and ancient people; and Reza Shah, who had seized power in 1921, was a fearsome monarch who—even though originally a mere trooper—soon acquired the traditional splendour and mystique of the Peacock Throne. In 1941, when Hitler invaded Russia and the Shah refused to expel his Nazi allies, the British and Russians invaded Iran to safeguard the oil and supply-routes. The Shah was exiled, first to Mauritius and then to South Africa, where he later died. The country was then ruled by the British and the Russians until after the war, when the Russians were with difficulty pushed out of the north.

In the old Shah's place the British put his twenty-one-year-old son, then a slight, inexperienced youth who had been educated in Switzerland and had grown up in awe of his father. He was intended as a puppet, and in Churchill's words: 'We have chased a dictator into exile, and installed a constitutional sovereign pledged to a whole catalogue of long-delayed, seriously minded reforms and reparations.' But that was not how the new Shah saw it. As he put it to me thirty years later: 'We were an independent country, and then all of a sudden the Russians invaded our country and you British took my father into exile. Then we were hearing that the oil company was creating puppets—people just clicking their heels to the orders of the oil company—so it was becoming in our eyes a kind of monster—almost a kind of government within the Iranian government.'[5]

The fact that the oil concession was in the hands of a single British company made BP an easy scapegoat for anything that went wrong in the country. The mood of the Iranian parliament, the Majlis, was increasingly nationalistic in the post-war years, and the resentments became greater when the British Labour government limited all company dividends, thus cutting back the Iranian budget. Sir William Fraser when he visited Iran was aware that concessions must be made, and by 1949 BP had devised a 'supplemental agreement' which at the time offered a rather better deal to Iran than other Middle East countries.

But the Majlis refused to discuss it, and soon afterwards the 'fifty-fifty' deals began to be spread to the Middle East, first to Saudi Arabia, then to Iraq. Aramco insisted that they had warned Sir William of this impending precedent, but if so, he did little about it. In fact, the supplemental agreement was as generous as Aramco's, allowing for the subsequent Aramco discounts, but the simple slogan of fifty-fifty provided a new rallying-cry for the Iranians. The militancy was encouraged by an outspoken American Ambassador, Henry Grady, a first-generation Irish-American who did not conceal his hatred for British imperialism[6] and who encouraged the Iranians—quite misleadingly as it turned out —to believe that the Americans would support them against the British.

And in the meantime an exotic new Iranian leader had emerged, Dr. Mossadeq. To the British public and the press this wild old man wearing pyjamas and perpetually weeping appeared ridiculous, so fanatical and unashamedly emotional, that he represented the defiance of all reason. But he was in fact a shrewd politician from a rich landowning family, who perceived the nationalist mood. Since the ill-fated Mexican nationalisations of 1938, he was the first leader to dare to confront the oil companies. Mossadeq was appointed chairman of a committee on Iranian oil policy, which insisted that BP's agreement did not safeguard Iranian interests, and became more militant after the announcement of Aramco's fifty-fifty deal in Saudi Arabia. By February 1951 Mossadeq was calling for nationalisation. Events then moved swiftly. The prime minister, General Razmara, insisted that Iran could not legally repudiate the concession. Four days later he was shot dead on his way to a mosque. Six weeks later, after strikes and martial law, the Majlis elected Mossadeq prime minister and voted for the immediate seizure of BP's oilfields.

The Iranian crisis thus suddenly became a challenge to Britain's world authority—in some ways more critical and unexpected than the Suez crisis five years later. The British Labour government were abruptly faced with the awkward question (which the Admiralty had raised forty years earlier, before Churchill had caused the government to buy its half-share of BP): should they intervene militarily to protect their investment? The argument that followed, still partly shrouded in secrecy, illuminated some of the fundamental problems of oil diplomacy. And it caused special embarrassment to a Labour government which had only recently itself nationalised major industries. ('We would never have dared to nationalise oil,' an Iranian official told Eric Drake, the general manager at Abadan and the present chairman, 'if Churchill had still been prime minister.')

The Labour government faced the crisis without much warning, for their information from Iran was scanty, and largely dependent on BP. The Foreign Office was in some disarray, with a newly-arrived Foreign Secretary, Herbert Morrison. Though he had been a pacifist in the First World War, Morrison was now hawkish, full of admiration for his Victorian predecessor, Lord Palmerston, and indignant about the Iranian's ingratitude. His militancy was encouraged by Sir William Fraser of BP and also by the Minister of Defence, Emmanuel Shinwell, and by many of the Admirals and Generals. But there was a notable exception. Lord Mountbatten, who was at that time Fourth Sea Lord, was convinced that intervention would be disastrous for Britain's whole future in Asia. He had negotiated India's independence as the last Viceroy and so he carried special authority with the government.

Most officials in the Foreign Office, too, were against intervention, largely on the grounds that it would be impossible to work the oilfields without the support of the Iranians, surrounded by a hostile country. Their position was supported by the Minister of State, Kenneth Younger. The Secretary for War, John Strachey, was also very sceptical of the need for force, and of the crucial importance of Iran. He was advised that the dispute must be settled quickly, either through force or otherwise, for three reasons. First Iran would be ruined, secondly both BP and the British economy would be crippled for lack of oil and thirdly, the Iranians could never themselves work the oilfields or the refinery. All three assumptions proved quite untrue.[7]

The cabinet were in some confusion, caught between conflicting advice from the military. A cruiser was sent out, HMS *Mauritius*, to patrol off the coast, and a parachute brigade was prepared. Morrison made a bellicose speech in parliament, but in the office he dithered. As one member of the Defence Committee now recalls it, 'it was a classic example of drift: it made me realise how governments can simply refuse to face up to contingencies.' Eventually on September 27 there was a crucial cabinet meeting: Herbert Morrison had returned that morning on the *Queen Mary* from the United States, where President Truman had made clear that he would not support the use of force. Morrison was still militant, and he was briefed on his way up from Southampton by Sir Roger Makins, a key diplomat at the Foreign Office. In cabinet, with the chiefs of staff present, Morrison expounded the case for intervention but he was cut short by Attlee, the Prime Minister, who had been concerned with Iran in Morrison's absence. Attlee insisted that force was quite out of the question, and no other member of the cabinet cared to challenge him. The government and BP thereafter prepared themselves for a more patient and subtle strategy.

In the meantime all attempts at negotiation had failed: the Iranians had formally taken over the great refinery at Abadan, all exports had stopped, and the British officials were harassed and impeded. Eventually in October 1951, the last remaining British employees and wives gathered at the Gymkhana Club, and embarked on the *Mauritius*, which steamed slowly up the river while the ship's band played Colonel Bogey.[8]

The more liberal members of the Foreign Office were exasperated to discover that they had been trapped by the narrow-mindedness of Sir William: BP in fact had become a Frankenstein monster, beyond control of its makers. This is how Kenneth Younger summed up the situation (in a hitherto unpublished memo to Herbert Morrison, the Foreign Secretary) in October 1951, complaining about BP, *alias* Anglo-Iranian:

> The principal reason why our advance information was inadequate was the short-sightedness and the lack of political awareness shown by the Anglo-Iranian Oil Company. They were far better placed than anybody else to make a proper estimate of the situation but, as far as I am aware, they never even seriously tried to do so. Sir William Fraser is no doubt a very good business man in the narrow sense, but on every occasion when I have seen him, either at Ministerial Meetings or elsewhere during these months, he has struck me as a thoroughly second-rate intellect and personality. He has on many occasions explicitly stated in my presence that he does not think politics concern him at all. He appears to have all the contempt of a Glasgow accountant for anything which cannot be shown on a balance sheet. This is an attitude quite incompatible with the responsibilities of the head of a company like A.I.O.C. operating in so complex and unsettled an area as the Middle East.
>
> It may well be that the Supplemental Oil Agreement was quite

a reasonable proposal when it was put forward and it may in practice be as favourable to the Persians as a fifty-fifty arrangement. There can, however, be no doubt that it was drawn up in a manner which makes it seem far less favourable than the Aramco Agreement. It is quite astonishing to me that when the Aramco Agreement was published even so limited a man as Sir William Fraser did not immediately see the writing on the wall . . .

This criticism of the A.I.O C does not, however, absolve the government from blame. H.M.G. hold 51 percent of the shares and nominate two directors. There is therefore no excuse for the government not having ensured adequate political direction of the company . . Directorships have normally been used to give rewards for superannuated public servants, and once appointed they have not been expected to take a very active part in the affairs of the company. Indeed had they done so it would have been regarded as undue interference with a business operation. The two directors in the present instance are in any case not the sort of people who would have been appointed if the posts had been thought to carry with them major political responsibilities.

Sir William Fraser, however, was still firmly in command. He was confident that before long Dr. Mossadeq would topple, and the Iranians would be forced to negotiate. For oil was far their biggest export—far more profitable than caviar or carpets—and their oil could no longer be sold. Early in the crisis BP had enlisted the support of the other six sisters, to make sure they would not buy 'hot oil' when it was nationalised, and on May 15, 1951, the State Department announced that the American oil companies had indicated 'that they would not, in the face of unilateral action by Iran against the British company, be willing to undertake operations in that country'. And BP were soon, with government help, able to enforce the boycott. When a Panamanian ship, the *Rose Mary*, took on oil from Abadan, RAF planes forced it into Aden harbour, where its cargo was impounded. BP's attitude was supported first by the Labour government and then by Churchill's Conservative government which came to power in October 1951.

BP's position was helped because there was at the time no serious shortage of oil. BP quickly used its huge reserves of capital, conserved under Fraser's cautious regime, to expedite its production in other countries, particularly in Kuwait, which soon began to make up for the loss of Iran. The American companies flew out extra production equipment to Saudi Arabia and Kuwait. Each of the seven sisters had an interest in proving to the Iranians and to other potential miscreants that they could quite well do without their oil. Iran faced the bleak prospect of being permanently left out of the world's oil markets.

ACHESON v. ANTI-TRUST

To the American government, however, the situation looked different, and from the beginning President Truman was concerned about Britain's stubborn attitude. He warned Britain of the dangers of using force and had tried to mediate through a visit by Averell Harriman to Teheran. The Secretary of State, Dean Acheson, though theatrically Anglophile, was appalled by the uncompromising imperialism of the British, and when Mossadeq came to America to argue his case to the U.N. Security Council Acheson struck up a kind of comic friendship with the emotional old man, fascinated by his pixie quality and his bird-like movements.

Acheson had little sympathy for BP's predicament; 'their own folly had brought them to their present fix', he wrote later 'which Aramco had avoided by (in Burke's phrase) graciously granting what it no longer had the power to withhold.' He thought that the BP bureaucracy had poisoned the judgment of the British government, and committed them to 'rule-or-ruin'. He was also much more worried by the possibility of a Communist takeover, and not at all averse to a stronger American presence in Iran.[9] As the British continued their long boycott of Iranian oil, Acheson decided in October 1952 that the Americans should try an independent initiative. But it could only be done with the help of the American oil companies, who alone had the tankers to take oil from Iran. As in their tax arrangements with Aramco the previous year, the State Department were only too glad to use the oil companies as part of their bulwark against Communism.

But the timing was specially awkward, for it was now that the engines of anti-trust were stirring again. While the crisis was developing in Iran, a major conflict began to be fought out secretly in Washington, whose details have only recently come to light.[10] This conflict illustrated further basic dilemmas of oil policy: caught between the democratic principles of anti-trust and the imperatives of foreign policy; between the long-term strategy of governments and the short-term tactics of companies; and between the widening differences of British and American attitudes to the Middle East. The battles of words in Washington provided a fascinating counterpoint to the diplomatic battles between Iran and Britain.

The anti-trust mole had been dormant since Pearl Harbor. Thurman Arnold's anti-trust crusade, while noisy and well-mounted, had achieved very little beyond the humiliation of Teagle and Exxon. But the trust-busters had now acquired more staff and expertise, and they were given new boost by the national defence hearings after the war, which had unearthed the damaging facts about how Aramco overcharged the Navy, and by the disclosures of overcharging from the ECA. The Justice Department still had the subpoenas and documents that they had collected in 1940 and '41, and they began searching the company files again in 1946 and '47.

In December 1949 the Federal Trade Commission resolved to investigate the foreign agreements of the oil companies, and in October 1951 their staff produced a report of historic importance. It was not published till a year later, in August 1952 (with some excisions about Iraq for national security which were only revealed twenty years later).[11] It was released not by the FTC but by the Senate Select Committee on Small Businesses, whose chairman was Senator John Sparkman. Sparkman wrote a stirring introduction, strikingly bold compared to the Sparkman who is today chairman of the Foreign Relations Committee: 'Practically alone among the great nations of the world', wrote the young Senator, 'the United States—through fact-finding and through enforcement of its anti-trust laws—endeavours to hold in check the power of giant organizations.'

The report thus made public was a political bombshell, and its title alone was a red rag to the companies, for it was called simply 'The International Petroleum Cartel'. It provided the first full account of the cartel agreements of the seven sisters, going back to the Achnacarry Agreement of 1928: it published the details of the Red Line Agreement, and analysed in detail the intricate sharing arrangements throughout the world. It was largely put together by a radical economist, John M. Blair (now Professor at the University of South Florida) who had painstakingly pursued the threads of evidence through the company files. The green paper-bound book of 400 pages (though long since out of print) has provided an indispensable source for critics of the oil industry ever since.

The gist of the FTC findings was that the seven companies controlled all the principal oil-producing areas outside the United States, all foreign refineries, patents and refinery technology: that they divided the world markets between them, and shared pipelines and tankers throughout the world; and that they maintained artificially high prices for oil. These formidable charges were now taken up by the Justice Department, determined to mount a new anti-trust case, at the precise moment when Dean Acheson was preparing his initiative in Iran.

The subsequent arguments provided a fascinating glimpse into the ideological strains that oil policy had brought about. In April 1952 Acheson, writing to the anti-trust division, admitted that the State Department should not interfere with anti-trust laws, but pointed out that the case against the oil companies might seriously impair American foreign policy aims and would probably weaken the political stability in the Middle East. But the Justice Department nevertheless went ahead, six months before the end of Truman's Presidency ('the only way to bring an oil case is if the President is not running for re-election', explained one of the staff, Ken Harkins, years later).[12] On July 17 the Attorney-General, James McGranery, announced a grand jury investigation into the international oil cartel, and the next month subpoenas were issued against twenty-one of the oil companies.

Acheson was now busily preparing his plan to enlist the American oil companies in a joint operation to rescue the situation in Iran. Attorney-General McGranery pointed out that the two conflicting policies 'presented a most formidable legal meal', and the trust-busters opposed the proposed new consortium as an 'unreasonable restraint of trade'. Acheson on his side was increasingly exasperated by the 'police-dogs from anti-trust', who 'wanted no truck with the mammon of unrighteousness.'[13] By January 1953, in the last days of the Truman administration, the two departments were at total loggerheads: and their two opposite viewpoints about the oil companies were both eloquently stated in reports to the National Security Council.

The first report, from the Department of State, backed by Defence and the Interior, spelt out boldly the case for strong government support of the sisters. The companies, it said 'play a vital role in supplying one of the free world's most essential commodities' and it boldly admitted: 'American oil operations are, for all practical purposes, instruments of our foreign policy towards these countries'. Moreover, the report continued, the oil companies played a significant role in the struggle of ideas with the Soviet Union so that 'we cannot afford to leave unchallenged the assertions that these companies are engaged in a criminal conspiracy for the purpose of predatory exploration'. The FTC Report had already been given regrettable publicity abroad, particularly in Venezuela and Britain, thus endangering the national interests of the United States; in Venezuela, the State Department reported, the oil companies' programmes had been set back years by the allegations. A criminal anti-trust action against the oil companies would be far more damaging, encouraging other countries to believe that 'capitalism is synonymous with predatory exploitation'. The criminal action against the oil companies should be abandoned in favour of a civil action, which could avoid a trial and attendant publicity. And a new commission should study the inter-relationships of anti-trust, security and foreign policy, to ensure a co-ordinated policy towards the oil companies.

The Attorney General wrote an equally incisive report arguing the opposite case: that the seven sisters, far from being crucial for national security, were profoundly damaging to it.

It is imperative that petroleum resources be freed from monopoly control by the few and be restored to free competitive private enterprise . . . Free private enterprise can be preserved only by safeguarding it from excess of power, governmental and private . . . The world petroleum cartel is an authoritarian, dominating power over a great vital world industry, in private hands . . . a decision at this time to terminate the pending investigation would be regarded by the world as a confession that our abhorrence of monopoly and restrictive cartel activities does not extend to the world's most important single industry.[14]

It was a new form of an old argument: how far should democratic institutions be abandoned in the defence of a democracy? Each side had telling points. It was true, as the State Department insisted, that the FTC revelations about the oil cartel were already encouraging producing countries to take bolder action against the companies. It was also true, as the Justice Department proclaimed, that a monopoly of oil production in the hands of seven companies did not square at all with the principles of a competitive, free-enterprise democracy. But between the two extreme positions, of either totally supporting the cluster of companies, or breaking them up, neither side really faced the problem, what the relationship between the companies and government should be. Yet both sides tacitly assumed that oil was far too important to be left to the oilmen.

It was Acheson's arguments, of course, that won the day. Six days after the reports were submitted, President Truman wrote to his Attorney-General saying that the interest of national security might best be served by changing the criminal case against the companies into a civil case—with the proviso (soon abandoned) that the companies must agree to subpoenas. And soon afterwards, Eisenhower was President and Dulles was Secretary of State. There was now little doubt in Washington that the global battle against Communism must take precedence over any complications about anti-trust. Nevertheless, the civil action against the companies went ahead, with a warning from Dulles that proceedings should be conducted 'with due regard for all matters affecting the national security'. On April 21 the historic complaint was filed, against five defendants (Shell and BP being beyond the government's jurisdiction): the United States versus Exxon, Mobil, Socal, Texaco and Gulf. It charged them, on the evidence of the FTC report, with perpetuating the 'As Is' cartel that had been established twenty years before, in a world-wide pattern of control of foreign production, refining, patents, pipelines and tankers: 'The objective of the conspiracy was market stabilisation; its essential terms were market division and price fixing.'[15]

COUP + CO-OPERATION

In the meantime Sir William Fraser and the British government had been waiting for nearly two years for the expected collapse of Mossadeq. Eisenhower and Dulles were having little more success in their negotiations than Truman and Acheson. Eventually in May 1953 Mossadeq wrote to Eisenhower complaining that his country was being ruined by the political intrigue of the British government and BP, and appealing for aid. Eisenhower did not reply for a month—which thoroughly unnerved Mossadeq—and then wrote back insisting that Iran must live up to her international obligations. Mossadeq was desperate; his nationalised oil company

still could not sell anything abroad, and there was now growing un-
rest in the country.

Behind the scenes mysterious forces were at work in Iran, who
were waiting for their moment. Early in the crisis, British secret
agents had reported to London that there were many anti-
Mossadeq elements in Iran who with encouragement, including
cash, from Britain, could help bring Mossadeq down. The
Foreign Secretary, Anthony Eden, however, would not sanction a
coup, and the project was passed on to the CIA in Washington,
who were in turn hesitant to act without British support. Eventually the plan was sanctioned, not by Eden, but by Churchill, who
happened to be in temporary command of the Foreign Office
during Eden's illness in April 1953. The conspirators were duly
assisted, master-minded by Kermit Roosevelt, and their chance
soon came. Mossadeq took control over the army; the Shah tried
to oust him, failed, and fled the country in August. Three days
later the Shah's supporters, led by paid agents, took to the streets
to fight Mossadeq's troops. Mossadeq was forced out of office, the
Shah's man General Zahedi took over the government, and the
Shah returned to his capital in qualified triumph—conscious that
it was more the CIA's victory than his, though he did not care to
admit it.[16] The cost of the covert operation, according to one
source, was about $700,000.

Whether and when Mossadeq would have fallen without this
covert intervention is hard to establish; but what is undisputed is
that the Western powers did intervene, and hastened his end. It
was a well-organised coup, and encouraged the CIA to further
adventures, notably in Guatemala; but the West in the end paid a
heavy price for it. For the Shah was thereafter determined to show
his independence, and could never again dare to be seen as the
pawn of the West.

The way was now open for the entry of the oil companies, but
there was still great diplomatic concern, and a concerted Anglo-
American policy now seemed urgent, for the Iranian experience
had made both governments more aware of the vulnerability of
all the Middle East oilfields. But while there were great differences
within Washington about the role of the oil companies, there were
even greater differences across the Atlantic. The British looked at
the anti-trust campaign with incomprehension: why shouldn't
companies and governments work together to protect the consumer? The British government proposed joint talks about Middle

East oil, which were held between August and October 1953 in Washington.

The notes of the discussions, only recently declassified, reveal the extent of the cross-purposes. The British team, led by Harold Beeley (later Ambassador to Saudi Arabia and Egypt, and one of the ablest Middle East experts) was very conscious of Europe's dependence on Middle East oil: the problems, they insisted, were 'highly strategic and political, not just economic', and Britain was 'not primarily concerned in protecting the profit position of the British companies'. The British wanted coordination between companies and consuming government, and to improve relations with the oil producers to ensure a safe flow of oil. They even proposed, twenty years ahead of time, an International Petroleum Council made up of producing and consuming countries. But the State Department was lukewarm on all the proposals. They could not encourage joint company talks because of anti-trust; they were afraid that consumer-producer talks would only encourage Arab co-operation; and they preferred to keep aloof from their own companies, particularly Aramco, for their own diplomatic reasons. 'Certain advantages,' they said, 'had flowed from this separation of identity, particularly during the early days of the development of the Israel State.' By allowing Aramco to stand on its own feet the government could avoid a clash between its two conflicting foreign policies. The talks eventually fizzled out. The chance of any co-herent Western policy was lost, and any hopes that Britain could improve relations with the producers were dashed by the Suez adventure two years later.

THE CONSORTIUM

By the time of the fall of Mossadeq, Sir William Fraser was reluctantly realising that BP could no longer retain the monopoly of Iranian oil which they had enjoyed for forty years. The wave of resentment was too strong and the future security of Iran rested more with the Americans than the British. The diplomacy thus was largely in American hands. Dulles appointed a petroleum adviser, Herbert Hoover, the son of the ex-President who had already mediated in Mexico, and Hoover flew between Teheran, Washington and London, trying to arrange for Iranian oil to flow again into the world's markets.

But the problem was thorny. Hoover wanted to include any

American oil company big enough to develop the Iranian oilfields —which meant all the five American sisters. But the companies were not very keen, since there was now again a world glut, and any new oil would mean cutting back oil from other countries. And for the five companies to march in together would make a total mockery of the anti-trust case that was brewing up: it would put the companies in a strong position to insist that the case be dropped altogether. But the trust-busters were now in retreat. In August 1953, Eisenhower told his Attorney-General to 'develop a solution which would protect the interests of the free world in the Near East as a vital source of petroleum supplies', and stressed that the enforcement of anti-trust 'may be deemed secondary to the national security interest'.

There were more spikey arguments between the State Department and the trust-busters; Leonard Emmerglick, the attorney who was handling the case, warned that the anti-trust division could not allow the five companies to co-operate if there were to be 'a regime of agreed-upon cutbacks or artificial controls' in the Middle East.[17] His concern was well-based; for such a systematic restraint of production went to the heart of anti-trust principles. But the choice had already been decisively taken by the State Department, whatever the case against the 'so-called cartel' (as they referred to it). In the cause of defence, and in the fight against Communism, the five sisters must be brought into Iran.

Hoover gradually paved the way to a coming-together of the companies. In December 1953 Sir William wrote to the five American chief executives: Holman of Exxon, Jennings of Mobil, Follis of Socal, Leach of Texaco, Swensrud of Gulf; and also to Shell and CFP. To each he explained that 'the companies which could make a constructive contribution to a solution are those which are now engaged in the production of oil in the Middle East and in the marketing of it on a large scale internationally', and he invited each to a meeting in London to discuss the Iranian problem. Exxon, as the senior company, passed the invitation to the State Department, asking whether they had any objection, and piously stressing that the company was very conscious of the 'large national security interests'. The State Department agreed, provided Herbert Hoover was present at the discussions.

Texaco, when they got the letter, shrewdly went direct to the Justice Department, and the anti-trust chief, Stanley Barnes, was

immediately worried that approval would 'make it impossible to
go on with the cartel case'. It was unfortunate, Barnes commented,
'that Hoover looked only to the cartel companies.'[18] The anti-
trust difficulty soon came to a head. The trust-busters continued
protesting, particularly about the lack of price competition, but on
January 14 the National Security Council gave its support to an
international consortium: and the Attorney-General then told
Eisenhower that in view of the security requirements the consortium 'would not in itself constitute a violation of the anti-trust
laws'.

The seven companies, in the meantime, had met in London
with Sir William Fraser and Herbert Hoover, together with the
French Compagnie Française des Pétroles, which was now a more
formidable power in world oil, regarded as a wayward eighth
sister. The eight each agreed to join a consortium to buy and
develop Iranian oil.

CONCILIATION

There remained the problem of squaring the Iranians, which
even after the fall of Mossadeq was not easy. The Shah, conscious
of the indignity of his dependence on the CIA, was determined
to drive a tough bargain, and Sir William Fraser was still very
stubborn. In December 1953 the Foreign Office had resumed
diplomatic relations with Iran, and sent out an open minded
young *chargé*, Denis Wright. Fraser was outspokenly critical of the
appointment, and regarded himself as the maker of diplomacy,
while the politicians became increasingly exasperated by his
stubbornness: 'Why don't we make him a peer and be done with
him,' said Harold Macmillan, a member of the cabinet. BP was not
now a buffer but an obstruction. Eventually Fraser was persuaded
that it would be impossible for BP to lead the negotiations; and
the task of heading the team was given to Exxon, while BP sent
out a patient mathematician, Harold Snow, who was regarded as
too silent to give offence. But the Foreign Office were behind the
scenes, virtually supervising the critical negotiations.

Exxon eventually brought in as their negotiator Howard Page,
a man who was to have an enduring influence on the Middle East
over the next fifteen years, as the master-diplomat among the
American oilmen.

Surveying the Iranian situation, Page was confident that a deal
could be done, supported by Wright. Page flew out to Teheran,

with Snow of BP and John Loudon of Shell, and began talks with the Iranian finance minister, Ali Amini. It was an exasperating haggle; the Iranians were using the Russians as a bargaining counter, and were determined to appear to drive a hard bargain. But in fact the Shah needed a settlement as much as the Westerners; and at last in August 1954 an agreement was signed by the heads of the companies, and was soon afterwards approved by the Majlis.

This new carve-up included for the first time all seven sisters and seemed at the time satisfactory to all sides. BP would take forty percent of the shares in the Consortium: the five American sisters would each take eight percent: Shell would take fourteen percent; and CFP would take six percent. The National Iranian Oil Company, which Mossadeq had created, would remain the owners of the oilfields and the refinery, and the Consortium would buy the oil from them. But through their exclusive control of the markets, the Consortium were now the effective masters of the oil-production. And BP's demands for compensation were satisfied, after long negotiations between Amini and Harold Snow. BP would be paid both directly from the Iranians and indirectly through payments by the other Consortium members. In the end BP got more than the Iranians had originally offered, and Sir William, to the anger of the British Treasury, got an extra tax benefit from the British government.

But there was also a more clandestine understanding between the companies; a 'participants agreement' which was signed by the eight companies and kept secret from the public and from the Iranian government for the following twenty years; and with special reason. For it described not only the terms under which they would buy oil but also how they would restrict production to avoid a glut. This was achieved by the formula of the 'Aggregate Programmed Quantity', or APQ, which was outlined in the Participants agreement, which was really to be the arbiter of Iran's future growth. The APQ calculated the total amount of oil that was to be 'lifted' from Iran in the following year, and it was reckoned by listing the needs of each participant, divided by their percentage share in the Consortium, in order of magnitude; and then taking the last figure after seventy percent of the holdings had been listed. A company wishing to take more than its quota would have to pay more for it. To take an example, in 1966 the APQ came to 1,945,000 barrels a day, calculated thus:[19]

Participants	Share (%)	Cumulative Shares (%)	1,000 barrels per day required
Independents (Iricon)	5	5	2,030
BP	40	45	2,027
Shell	14	59	2,027
Mobil	7	66	1,964
CFP	6	72	1,945
Exxon	7	79	1,890
Texaco	7	86	1,712
Gulf	7	93	1,700
Socal	7	100	1,644

The significance of this system was that it effectively held down production in Iran to the levels required by the least demanding of the companies. If Exxon and Texaco for instance, were to want less oil (as they always did) because of their commitments in Saudi Arabia and elsewhere, BP and Shell would have to restrict their production, too. It was not surprising that the companies did not wish the Shah to know how his country's future income depended on a private rationing system controlled by eight companies. When he *did* know, as we will see, the effect was electric.[20]

The secret clause had to be revealed to the anti-trust staff in Washington who were preparing their civil case against the cartel and it bore out their worst fears. Leonard Emmerglick, in charge of the case, warned that the Attorney-General would have to 'stultify himself' to approve the arrangements, and Kenneth Harkins, his number two, believed that they 'manifest a continuation of the cartel pattern'. The Attorney-General, Herbert Brownell, agreed that approval of the Consortium was inconsistent with the cartel case, and admitted that the case must proceed with emphasis on the marketing aspects and not on the production control aspects'. But he repeated his assurance to President Eisenhower that in view of the security interests, the Consortium would not violate the anti-trust laws.

The smaller American oil companies not surprisingly complained about being left out of this cosy cartel, and eventually in April 1955 a concession was made to them: each of the five American participants gave up an eighth of their holding, to allow five percent of the Consortium to be held by nine of these 'independents'.[21] But the method of choosing them aroused the suspicions of the trust-busters, for they had all to be approved by the accounting firm of Price Waterhouse, which for some time had been the special vehicle for the American sisters in their joint operations.[22] One company not so approved, American Independent, had to appeal to the State Department to be allowed onto the list. The enlargement of the Consortium also worried the Texan oilmen, who were fearful that they would be flooded by cheap Iranian oil. Their chief defender in Congress, Senator Lyndon Johnson, passed on their concern to Dulles, and was then assured that the companies 'could and would absorb the Iranian production without unsettling world markets'.[23]

But the Consortium had a far more serious limitation in global terms: for it represented companies from only four consuming countries. The three defeated nations whose consumption of oil was now rapidly growing—Italy, Germany and Japan—were unrepresented in the share-out. The exclusion of Italy, the most vocal of them, was to have swift repercussions, as we will see in the next chapter.

As for the case against the cartel, it continued to be assembled over the following years, occupying huge staffs of attorneys in the oil companies and in the Justice Department. It was not until 1961 that a 'Statement of Claims' was put forward against the five sisters, signed by Lee Loevinger, the anti-trust chief under the Kennedy administration. It was, like the complaint of eight years before, boldly worded: 'The fact that the conspirators were able to control market prices and prevent surplus oil from upsetting the market equilibrium', it concluded, 'demonstrates the existence of monopoly power'. But as a result of the Iranian Consortium the statement concentrated much less than the previous complaint on the joint production agreements, which were the heart of the collusion, which took much force out of the charge. Eventually three of the companies—Exxon, Texaco and Gulf—agreed to consent decrees, and the two other cases, against Mobil and Socal, were later dismissed without prejudice. The trust-busters, for the time being, had lost ground.

SEVEN IN CONTROL

By October 1954 the oil was again flowing from Iran to the world's markets, and technicians were returning to restore and enlarge the oilfields. The three-year crisis was over; yet what was remarkable, after the first panic, was how little of a crisis it had been. The oil companies had shown first that they could do without Iranian oil, and then that they could again do with it; and neither the turning-off nor the turning-on had any perceptible connection with the price of oil. It was a phenomenon that gave much support to the suspicion that the seven if not actually a cartel, worked in some sort of close unison, keeping down the supplies of oil to keep the price up, and co-ordinating supplies throughout the world. As one prominent student of the companies concludes: 'It seems incontestable that the rate of supply was subject to some sort of controlled planning . . . but there is no evidence at all that control was exercised in any *centrally* co-ordinated manner.'[24] This was the central mystery behind the symbiosis of the seven.

BP, while losing their monopoly, had achieved handsome compensation and a promising future; their shares had zoomed up from £5 in October 1951 to £18 after the consortium agreement was signed, and in November 1954 a bonus of £80 million was distributed to shareholders. In fact BP men were later able to reflect that the whole crisis was a blessing in disguise; for the temporary loss of the oilfields was a shock which forced them rapidly to find and develop oilfields elsewhere, and they would never again be so dependent on a single country. It was a shock that has been compared to the shock to Standard Oil at the dissolution of 1911.[25] And perhaps some kind of crisis would have occurred anyway before long, even without the intransigence of Mossadeq and Fraser: 'The tide of nationalism was already flowing', said Sir Eric Drake; 'it could only have been held back for a few more years.' The all-British monopoly would have been hard to maintain, and the transition to the Consortium reflected the new realities of American power. It had been achieved with tact and dignity: BP were able to cover up their humiliation with a show of dignity: Sir William, who was a prime cause of the disaster, was acclaimed by the British press as the architect of the settlement (with scarcely any reference to the Americans' role) and the next year was created Baron Strathalmond of Pumpherstone (he should, commented Gulbenkian, have been called Lord Crude of Abadan).

Moreover, all the seven sisters could congratulate themselves that the Iranians, and all other potential nationalisers, had been taught a lesson. They could not do without the big companies, as the Mexicans had found thirteen years before: they could only drown in their own oil. The companies had again made clear that they controlled the world's markets: and the American five, though still harassed by anti-trust, had gained greater recognition from their government. For the next twenty years, they could warn nationalist leaders, explicitly or implicitly, 'look what happened to Mossadeq': and for at least sixteen years—so long as there was a large surplus of oil—the warning was effective.

And the seven had now achieved a more perfect system of control and orchestration of the supplies of Middle East Oil: the Iranian Consortium, unlike the consortia in Iraq or Saudi Arabia, included all of them, and this time they had brought in the French company too, thus avoiding the Gallic acrimony attending the Aramco agreement. In each of the major producing countries, two or more of the seven were in control, and all four of the partners of Aramco were now also partners in Iran, making much smoother the delicate task of balancing production between them. The chart overleaf shows the share-out between the major countries from 1954 onwards.

From the viewpoint of the seven there was nothing sinister about the carve-up: it merely reflected the elementary fact that only huge companies with resources and markets were in a position to develop and sell the oil. And without joint arrangements, in a world of glut, there would be only cut-throat competition and sudden floods of cheap oil, such as had made life so intolerable in the mid-'twenties: it was only common sense to agree to restrict supplies.

But to anyone outside the seven—including the Italians, Japanese and Germans, the American independents, and the producing governments themselves—the system was thoroughly pernicious: if they had not known it already, the FTC Report now explained it. For they saw the production of oil being manipulated by the seven to achieve the highest possible profits; and even with the lower price of oil, their profits were still enormous.

As for Dr. Mossadeq, his role in history is still disputed. Among old Iranians, he is now still an embarrassing phenomenon, who bankrupted his country and looked foolish to the world; and the Shah, who had to leave the country because of him, prefers not to hear the name of 'that fellow'. But to most younger Iranians, he is a kind of national hero, for he first asserted Iranian nationalism against the Company and the British. Even if his timing was wrong—as it manifestly was—his ideas still lived on. The Shah in

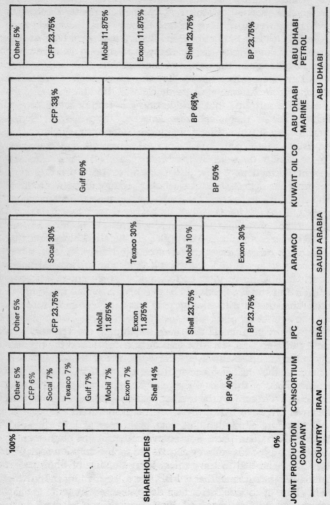

COMPOSITION OF THE KEY JOINT OIL PRODUCTION COMPANIES IN THE MIDDLE EAST 1972

JOINT PRODUCTION COMPANY	CONSORTIUM	IPC	ARAMCO	KUWAIT OIL CO	ABU DHABI MARINE	ABU DHABI PETROL
COUNTRY	IRAN	IRAQ	SAUDI ARABIA			ABU DHABI

SHAREHOLDERS

CONSORTIUM (IRAN): Other 5%, CFP 6%, Socal 7%, Texaco 7%, Gulf 7%, Mobil 7%, Exxon 7%, Shell 14%, BP 40%

IPC (IRAQ): Other 5%, CFP 23.75%, Mobil 11.875%, Exxon 11.875%, Shell 23.75%, BP 23.75%

ARAMCO (SAUDI ARABIA): Socal 30%, Texaco 30%, Mobil 10%, Exxon 30%

KUWAIT OIL CO: Gulf 50%, BP 50%

ABU DHABI MARINE: CFP 33%, BP 66%

ABU DHABI PETROL: Other 5%, CFP 23.75%, Mobil 11.875%, Exxon 11.875%, Shell 23.75%, BP 23.75%

Source:— Multinational Hearings Part 5

fact was deeply influenced ever after by his experience. Ever since he has had to show the Iranians that he can hold his own against the companies and the West; and while mocking Mossadeq he has tried to take some of Mossadeq's clothes.

Iran had been brought to heel with the acquiescence of other producing countries, who were happy to benefit from the boycott, and this was to be the source of some shame to them ever afterwards. When, six years later, they came together to form OPEC, the member-countries each made clear to the Secretary-General, Fuad Rouhani, that they would not take such advantage of rivals again. The lean spectre of the fanatical doctor flitted over the first counter-attack by the producing countries.

SUEZ AND AFTER

The formation of the Iranian Consortium marked in some respects the apogee of the influence of the seven sisters, both with the Middle East governments and with their own home governments. In Washington, the State Department was only too glad to let the companies look after the problems of oil, and to let Aramco pursue a separate foreign policy. In London, the Foreign Office was more sceptical of BP after the Iranian disaster, but the Treasury still regarded oil companies as geese laying golden eggs, in the form of huge contributions to the balance of payments: by 1956 BP and Shell were reckoned to supply £323 million a year[26] — more than half of Britain's receipts from all overseas investments.

The disastrous Suez adventure in 1956 seemed to corroborate the older view that oil *was* best left to the oil companies. Sir Anthony Eden launched the war in the name of protecting British interests, of which oil was far the greatest: two-thirds of the Suez Canal traffic was oil. The high emotions of Eden and his inner circle were certainly partly generated by a sense of panic about oil. As Hugh Thomas put it: 'Ever since Churchill converted the Navy to the use of oil in 1911, British politicians have seemed to have had a feeling about oil supplies comparable to the fear of castration.'[27] But there was no evidence of consultation beforehand with oil companies or experts before the invasion took place, and the heads of oil companies, John Loudon of Shell and the more enlightened members of BP, were appalled by the invasion. They saw it as wrecking their delicate relations with the Arab producers, with good reason. First the canal was blocked, so that tankers had to be chartered at exorbitant costs to take oil round the

Cape; then the IPC pipeline from Iraq to the Mediterranean was sabotaged by Syrians; then anti-British demonstrations broke out in Arab capitals and in Iran.

Suez in fact was a serious setback to the British companies, and further enhanced the American advantage in the Middle East. The American companies had been brought together with the government in a Middle East Emergency Committee (MEEC), to provide an 'oil lift' for Europe, pooling their resources and diverting oil to Europe. But while Britain was refusing to withdraw from Suez, the U.S. government deliberately suspended the committee as a means of pressure, and the British and French suspected that the American oil companies were using the crisis to fortify their own position.

Certainly they gained some commercial benefit: the Texan producers saw the temporary world shortage as the opportunity to push up the price, even though there was such huge excess capacity in Texas that production was cut to thirteen days a month under the pro-rationing agreement. Humble, the Exxon subsidiary, raised its price by 35 cents in January 1957, followed by others. There was an immediate outcry which led to Senate hearings on the effects of the emergency oil lift, in which Senator Kefauver was a prominent critic. The Justice Department investigated, and a federal grand jury indicted twenty-nine oil companies for conspiring to fix prices; but they were later acquitted by a District Judge in Oklahoma, Royce H. Savage, who explained: 'I have an absolute conviction personally that the defendants are not guilty.'[28] 'Why the federal government thus delegated the conduct of foreign policy to the Texas Rail Commission', comments Professor Adelman, 'will never be known . . . The price increases of 1957 marked the zenith of postwar control of the market.'[29]

The Suez fiasco provided new fuel for Arab nationalism, and made the oil companies more conscious of their role as diplomats and their advantages over governments. Two years after Suez, Shell were operating amicably in Egypt while the British government was unrecognised. And an Exxon memo in early 1957 described in statesmanlike terms how 'the problem of restoring confidence in Middle East oil is one which concerns every member of the petroleum industry active in Europe . . . We must be sure that we come out of the present emergency with undiminished prestige and strength.'[30]

NOTES

1. Dean Acheson: *Present at the Creation*, Signet, New York, 1974, p. 650.

2. John Strachey: *The End of Empire*, London, 1959, p. 173.

3. Henry Longhurst: *Adventure in Oil: The Story of British Petroleum*, London, 1959, p. 137.

4. Stephen Hemsley Longrigg: *Oil in the Middle East: Its Discovery and Development*, 3rd edition, London, 1968, p. 157.

5. Interview with author, February 4, 1975.

6. See article 'What Went Wrong in Iran?': *Saturday Evening Post*, January 5, 1952.

7. Strachey: p. 161.

8. Longhurst, p. 144.

9. Acheson, pp. 865–870.

10. Much new material about Iran is contained in the Sub-Committee print published by the Senate Sub-Committee on Multinational Corporations in February 1974: *The International Petroleum Cartel, the Iranian Consortium and U.S. National Security*, to which I am indebted.

11. Multinational Report: 1975, p. 101.

12. Mark J. Green: *The Closed Enterprise System* (Bantam Book), New York, 1972, p. 274.

13. Acheson: p. 869.

14. Subcommittee Print: 1974, p. 29–32.

15. Subcommittee Print: 1974, p. 125.

16. See his autobiography, *Mission for My Country*, London, 1961, pp. 99–105.

17. Subcommittee Print: 1974, p. 54.

18. Subcommittee Print: 1974, pp. 57–60.

19. Multinational Hearings: Part 7, p. 254.

20. In her book *The International Petroleum Industry* published in 1968, Dr. Edith Penrose reckoned from existing evidence that the cut-off point was between 60% and 79%, but the details of the agreement, and the actual figure of 70% were not published until 1974, when they made part of the evidence of the Senate Multinational Hearings.

21. See Chapter 7.

22. 'All through the documentary material delivered by the five defendants in the cartel case,' complained Watson Snyder of the anti-trust division in March 1955, 'you will find that Price Waterhouse and Company is the medium through which all the accounting is done for the participants in the various illegal arrangements.' Multinational Hearings: Part 7, p. 249.

23. Multinational Hearings: Part 8.

24. Dr. Edith Penrose: *International Petroleum Industry*, London, 1968, p. 152.

25. Penrose: p. 114.

26. Andrew Shonfield: *British Economic Policy Since the War*, London, 1958.

27. Hugh Thomas: *Suez Affair*, London, 1970, p. 32.

28. John Bainbridge: *The Super-Americans*, London, 1962, pp. 86–7.

29. M. A. Adelman: *The World Petroleum Market*, p. 158–9.

30. Quoted in Robert Engler: *The Politics of Oil*, Chicago, 1961, p. 190.

7

The Intruders

When you're breaking in, you're sons of bitches: then you become a member of the Club. We hope to join as the Eighth Sister.

Leonard McCollom of Conoco, 1967

Oil still had a place for the loner. The giant companies were building up huge bureaucracies and planning global operations with unequalled resources. But calculated gambling was still the heart of the business. There were still individuals who were able to challenge the established powers, armed with the same lonely ambition as the first Rockefeller. And among the forces that changed the course of oil history in the fifties was the solitary millionaire from Oklahoma, Jean Paul Getty.

GETTY

Today, at the age of eighty-two, Getty remains a monument to individualism, as he loooks back with melancholy detachment on his sixty years in the industry. He lives alone in the great Elizabethan mansion of Sutton Place, surrounded by parkland on the edge of suburban London, and protected by extravagant defences. A heavy double gate on the main road guards the estate, with electronic controls, followed by a traffic light. Guard dogs roam across the park, and inside the walled garden is a cage full of lions. Inside, the dark house seems lifeless, except for the security guards in the passages.

Visiting him in 1975 I was shown into a chilly sitting-room hung with Dutch masters, and waited in silence, until an old man shuffled in, with his huge long face hanging loosely, and his thin mouth breaking into a faint smile. He talked slowly and huskily, each reply ending in a low cadence and silence. The deadpan face and stretched cheeks were strikingly reminiscent of the photographs of the ageing John D. Rockefeller; and the comparison was one which Getty liked to encourage. He was proud of having been visited by John D. Rockefeller III, who presented him with

Professor Nevin's biography of his grandfather. He keeps the volumes in his study, together with his own autobiography.

The significance of Getty is not so much the fact that, like Rockefeller, he became the richest man in the world. It is that he was the first individual to challenge the dominance of the seven sisters in the Middle East, thus helping to begin the erosion of their monopoly. He typified the spirit of the rising 'independents'.

Getty was born rich: his father was a lawyer who visited Oklahoma during the first oil boom and stayed to make a multi-million dollar fortune. His son was fascinated by the 'nuts and bolts' of the oil business and made his own million dollars by the age of twenty-three. He studied at Oxford, travelled widely and fancied himself as a playboy, but was always obsessed by the oil business. His lonely detachment made him able (as he described it to me[1]) to avoid getting lost in the trees, and to see the forest more clearly, as if from a plane; and his pursuit of a succession of wives never interfered with his pursuit of money.

By 1934, when he was only forty-one, Getty was strong enough to be able to bid for a medium-sized independent company, Tide-water, and thus found himself challenging the majors: both Exxon and John D. Rockefeller II, it turned out, held shares in the company, and Walter Teagle tried to stop Rockefeller selling out to the newcomer. But he was too late, and Getty was in control of an integrated company which he greatly expanded. But he remained a loner, running his company from wherever he happened to be, and he saw himself not as a very rich man, but as a poor company competing desperately with giants. 'Actually I've never felt rich,' he told the *New York Times*, 'because I've always been in a business where I was a moderate-sized fellow compared to Exxon, Shell, Gulf, Socal. I'm a small-sized fellow, a small-sized outfit, so I've never had delusions of grandeur.'[2]

By 1958 when his riches were suddenly revealed by *Fortune* Magazine, he was discovered to be running his empire from the Ritz in London, buying and selling oilfields with the help of a single overworked secretary. He never became part of any organ-ised oil world, and kept away from his independent rivals. He never met the other oil billionaire, H. L. Hunt, but he once passed a message to his son Bunker that he would 'gladly turn over the appellation of the richest man'. Hunt, in reply, thanked Getty and said he preferred to be 'one of the pack'.[3]

Getty lost his first chance to get into the Middle East in 1932, when he turned down a concession in Iraq, and he did not explore further. East Texas was selling oil at a few cents a barrel, and 'it seemed rather pointless to go as far as Saudi Arabia'. But he later regretted it, and after the war he realised that the future lay with the Middle East. As early as 1948 there was a new chance to break into the monopoly of Aramco, when the Neutral Zone between Saudi Arabia and Kuwait was opened to bids. On the Kuwait side, the concession was won by a syndicate of Independents called Aminoil, headed by a veteran oil diplomat, Ralph K. Davies, with Phillips Petroleum as the major shareholder. They made a spectacular bid, including a down payment of $7·5 million, which horrified the majors. Soon afterwards Getty began negotiating with King Ibn Saud for the concession on the Saudi Arabian side.

Getty was not inhibited, as the majors were, by existing agreements. 'The credo of the seven sisters,' as he put it, 'was that they couldn't do any better. I think that if they had been more flexible years ago, they might have been better off today, but they took rather a hard line.' Getty saw that the King would welcome an outsider, and he eventually agreed on terms which included not only a down payment of $9·5 million, but much greater royalties. It caused fury among the majors: as Getty said, 'I think it proved to the Arabs that up till now they weren't getting the best terms . . . I told Arab friends of mine that I now felt I never had to run down an alley any time I saw an Arab approaching, because I gave the Arabs very good terms right from the start,'[4]

These ventures seemed to many onlookers to be wildly extravagant, and for four years Getty and Aminoil continued drilling without success. But eventually oil was found and was soon producing as much as sixteen million barrels a year. Getty was effectively established in the new centre of gravity, with a new fortune. The fact that a smaller company, offering far better terms than the majors, could still make large profits was not lost on the Saudis. Getty had already helped to breach the high wall of the seven.

INDEPENDENTS

Getty was soon followed by others, for the seven sisters were making such visibly huge profits in the Middle East that they lured others to challenge them. By the 'fifties a group of 'Independents' were beginning to move abroad. The name independent is confusing,[5] but in the international field, the word is usually applied to mean any company other than the seven majors: that is, a company without a global system of production, transport, refining and marketing. Many of them have achieved a size and wealth which in any other industry would seem vast, so that within the United States they are regarded as majors.[6] Some of them were originally offshoots of the Rockefeller trust, such as Standard Oil of Indiana (Amoco), Standard Oil of Ohio (Sohio), Continental (Conoco), Atlantic or Richfield, later merged to make Atlantic Richfield (Arco).

But others have more flamboyant and recent origins, having been built up by individual tycoons who had got rich in the oil booms of the south-western states. Harold Layfayette Hunt made his first fortune by buying out the oilfields of Dad Joiner in East Texas in the 'thirties: by the Second World War he was able to boast that he controlled more oil than the Axis powers put together. His son, Bunker Hunt, inherited his father's obsessive ambition, and determined to make his own fortune, outside as well as inside America. Armand Hammer had first become a millionaire with a pencil factory in Soviet Russia, where he enjoyed the friendship of Lenin: in 1956 he had bought up a moribund West-Coast company called Occidental and resolved to build it up into an international giant. Many of the independents were controlled by rugged entrepreneurs of the old mould, defying the rule of bureaucracies and consortia, and by the 'fifties they were beginning to look to the world as their stamping-ground.

These independent tycoons had a paradoxical relationship with the chief executives of the sisters. The independents became some of the richest men in the world including at least two billionaires, while the chief executives of Exxon, Mobil or Shell have relatively tiny fortunes, and salaries of a mere few hundred thousand dollars. But the independents often saw themselves as struggling underdogs, confronting the corporate power of the giants, and were prone to conspiracy theories: to Hunt down in Texas the Standard Oil Companies still represented a Rockefeller empire, closely tied to the eastern banking establishment. The independents despised the corporate heads for their lack of daring, while the big-company men deplored the crudity of the independents who were giving oilmen, once again, a bad name.

The independents liked to depict the majors, who were importing oil from Arabia, as potential traitors: Russell B. Brown, the general counsel for their association, complained in 1955 that 'their motivations have taken an a new and foreign accent as they have become companies of the world rather than of this nation'.[7] The independents' interest was still firmly based on Texas, which in 1960 was producing 39 per cent of United States oil, and their triumph was the triumph of Texas. The oilmen were apt to assume that, because they were lucky enough to be sitting on oil, their wealth was the result of special wisdom and genius; so that the Texan arrogance became hard for others to bear. The new oil millionaires felt no need to disguise or apologise for their wealth; they dominated the annual meetings of the American Petroleum Institute with extravagant razzmatazz, turning the city upside down and boasting of their political clout.

They knew that Washington was obedient to their will. The late 'fifties were the heyday of the oil lobby, with two Texans in key positions: Lyndon Johnson as Senate Majority leader, Sam Rayburn as Speaker of the House, and a sympathetic Congressman from Arkansas, Wilbur Mills, as Chairman of the House Ways and Means Committee. They were all steadfast in their support of the oil interest, and particularly of the crucial tax-deduction, the 'depletion allowance'.

This remarkable law, first passed in 1926, allowed oil companies to set 27½ percent of their gross income (provided it did not exceed half their net income) against taxes, to take account of the wasting assets below the ground. Each Secretary of the Treasury tried to reduce the allowance, and each one failed; Rayburn and Johnson, said Jim Clark, the Houston oil historian, 'stood like Horatio at the bridge for years, defending depletion against all comers'.[8]

And the majors, in the meantime, had lost a good deal of their daring, both in America and Britain. In the post-war years there were no bosses comparable to Teagle and Deterding. There were some powerful characters who took critical decisions; men like Eugene Holman, who established Exxon in Arabia, or Sir William Fraser, who dominated BP, Gus Long, who was to dominate Texaco for a decade. But the new chief executives were nearly all men who had climbed their way steadily up through the companies, working through committees, and staying only a few years at the top. Monroe Rathbone, who took over Exxon from Holman, used to say that half his time was spent in thinking about personnel problems. The very extent of the big companies' commitments, and the range of their shareholding, had limited their decisiveness;

they were terrified that a bold move in one country would reper-
cuss on the others.

As they became threatened by outsiders, they became increasingly defensive, unable to accept change until it was forced upon
them.

> This present day desire to protect the status quo [wrote Richard
> Funkhouser, a prescient diplomat in the State Department in
> 1953], would seem to be a sign of old age in the business and does
> not reflect the attitude of the aggressive leaders who won the con-
> cessions in the first instance. If this is true, such companies as IPC
> and Aramco may look forward to the ignominious end of the
> unversatile dinosaur family that sank in the mud when the weather
> changed...[9]

The independents and the majors, behind all their outward
battles, realised that they could not do without each other, and
had reached an armed truce. The majors knew that the independents took risks in exploring which they would not take, and
provided them, too, with a kind of alibi against anti-trust action.
The independents used the majors to sell their oil, and needed
the ordered marketing system that they had set up. It was a
situation described in economists' shorthand as 'oligopoly with
a fringe': or more picturesquely as being like roses growing on a
pergola.[10] The major companies, having together constructed
their pergola, had an interest to maintain its stability, and they
could tolerate quite well the intrusion of the roses, which indeed
made the structure look more attractive.

It was the Iranian Consortium in 1954 which gave the first
opportunity for several of the independents to break into the
charmed circle of the seven, and to discover the Middle East. The
protests to Washington, as we saw in the last chapter, have compelled the big companies to include nine American outsiders. They
were Aminoil, Sohio, Atlantic and Richfield (later merged), Signal
and Hancock (later merged), San Jacinto (later bought by Continental) and Getty's two companies, Getty and Tidewater. These
newcomers soon received a wonderful bonanza from their access to
Iranian oil, which for them involved no risk or exploration. They
began to roam elsewhere in the Middle East for quick profits.

Their greatest opportunity came in Libya, where American
companies had been interested since the Second World War in
the possibilities of oil. After independence in 1951 the Libyan
Government was determined that their oil should not be mono-

polised by the majors, and the Petroleum Law of 1955 was drafted by a team of experts including Nadim Pachachi (later a Secretary-General of OPEC), who made certain that there would be a large number of concessions, and that those not exploited would have to be relinquished. In the following year, fifty-one concessions were granted to seventeen companies, led by a syndicate of three independents (Amerada, Continental and Marathon) in a group called Oasis. When the oil began to flow in large quantities, half of it came from the independents. Libyan oil gave these independents the chance to establish themselves in Europe and to challenge the marketing systems of the sisters. Perhaps the most remarkable of these was Continental (or Conoco), a medium-sized integrated company which was perilously short of crude oil. Libya provided it, and by the mid-'sixties, under a dynamic new chief executive, Leonard McCollom, Continental was already competing seriously with the majors. They bought their own distribution network—Jet in Britain, Sopi in Germany, and Seca in Belgium—to force their way direct into the markets.

Already by 1957 a glut of oil was beginning to force down the price. The independents could undercut the big companies and still make big profits for themselves. They competed more hectically to sell their Middle East oil to Europe and Japan as they began to be shut out of the United States. By 1957 the imports of oil to America had risen so sharply that the Texan producers complained of being done out of business; they invoked their old ally of 'national security', and demanded import quotas, The majors, led by Exxon, tried to resist, but by 1959 import controls were made mandatory, allowing only 13 percent of oil consumption to come from abroad. So the American companies had to sell most of their Middle East oil outside their home market, and the battle for Europe had the extra ferocity of a battle between exiles. The new flood of Libyan oil in the early sixties increased the pressure, and the sisters could not control it.

MATTEI

While the American independents were eating into some of the markets of the seven, a more far-reaching threat was emerging from a man who at first seemed merely comic: the head of the Italian State Oil Company, Enrico Mattei. He was combative, aggressive and exasperating; but the oil companies soon had to take him seriously, for he hit them on all their most vulnerable points. He was a nationalist, claiming that Italy had been done down by the 'oil cartel'. He was the ally of the producing countries, persuading them that they had been cheated. He was an individual tycoon, moving swiftly against the slow bureaucracies, yet with his own government behind him. And he was an arch-propagandist, presenting himself as the gallant little fighter against the seven sisters — *le sette sorelle* — a phrase which he rapidly popularised.

His personality was electric. I was invited to lunch by him in Rome in 1961 and his whole style was the opposite of the oilmen of Shell whom I had been recently interviewing. He was small and shy, with the look of an artist — a sensitive mouth, a mobile chin and a penetrating eye — but he visibly dominated his officials with an intimidating aloofness, and it was clear why he was called the 'incorruptible corrupter'. He talked with a flood of fast rhetoric: he carried on a separate conversation while his words to me were being translated.

He explained to me how *le sette sorelle* were relics from the colonial era, and how the producer countries no longer needed their technical know-how. The heads of the big companies were still making absurd, bullying speeches, but they would have to come to terms with nationalism; the Americans would do it first, as they had with the fifty-fifty agreement. Six years ago, he went on, he might have come to terms with the sisters, but now he was committed to helping the Arabs. It was clear that he was fascinated, perhaps too fascinated, by the mystique of the sisters and their power: and it was not quite clear whether he wanted to beat them, or join them.

Whatever else, he was a brilliant self-advertiser; and to every tourist in Italy, his publicity was then inescapable. The roads were full of bright shining filling-stations, under the sign of the six-legged dog which proclaimed Mattei's petrol, AGIP. In contrast to the austere stopping-places of Exxon or Shell, they had neat little bars and restaurants with bright-coloured tables, and cheerful assistants in clean uniforms. It all seemed the best kind of nationalism, a new pride in patriotic performance. Mattei made the seven sisters look not only mean and unloved, but boring and unimaginative.

Mattei's history was certainly more romantic than theirs. After being a wartime partisan leader, he was put in charge of the state-controlled oil company, AGIP. According to a much-publicised (but questionable) story, he was told to give up exploring in Italy but ignored his instructions, and soon afterwards AGIP discovered natural gas in the Po Valley.[11] Later they found a small quantity of oil — a thrilling development for a nation with so few natural resources—and the Italian discoveries, though they provided only a fraction of AGIP's oil, gave a special political glamour to Mattei's enterprise. By 1953 he had formed a new national hydrocarbon organisation called ENI and he was ready for battle with the seven sisters.

His first attempt to break into the ring came when the Iranian Consortium was formed in 1954. Mattei had behaved with studied retraint during the Iranian boycott, refusing to buy cut-price oil offered by Mossadeq; and he expected in return, as a major buyer of oil to be included in the Consortium. But the negotiators found him altogether too aggressive and impossible; and he was not invited—a rebuff which proved as unwise as the failure to invite the wicked fairy in the *Sleeping Beauty*. As Mattei's adviser, and later biographer, commented: 'It was almost grotesque to see that for no reason other than the desire to pay lip-service to free competition, a few American investors were, by courtesy of Washington, given what amounted almost to a free ride, although the oil they were to get played no functional role in their own activities, whereas a country like Italy and a refiner-distributor like AGIP were to remain outside.'[12]

Mattei's revenge came soon enough, in the wave of Middle East nationalism following the Suez War. He began talking to the Iranians about joint ventures with more genuine participation, and after the Iranians had passed a new Petroleum Law in 1957, he announced a sensational agreement with the National Iranian Oil Company to develop new concessions. They would be partners from the beginning, with an Iranian Chairman of the joint board: the Italians would pay for the exploration costs, and if oil was discovered the Iranians would take close to 75 percent of the proceeds, instead of the customary fifty-fifty.

This joint venture in fact found very little oil, but its political consequences were enormous. For Mattei had made clear to the Iranians, as Getty had shown the Saudis, that the seven sisters were making exorbitant profits; and the Shah freely acknowledges his debt to Mattei. Soon afterwards one of the big American independents, Standard Oil of Indiana (Amoco) followed Mattei into Iran, to make an even more favourable deal. And Mattei himself repeated his deal across North Africa.

Mattei quickly extended the Italian influence, exploiting the resentments of developing countries and exasperating the seven sisters. But he failed to make a profit out of it, for as the sisters pointed out, he was 'an oilman without oil'. Excluded from the preserves of the seven, he was constantly forced to look elsewhere. He eventually found some oil where so many previous outsiders — the Nobels in the 1890s, Samuel in the 1900s, Deterding in the 1920s — had found it before; in Russia. By bringing Russian oil back into Europe he helped further to undermine the solid foundations of the seven sisters.

The Russians in the late 'fifties, after big new discoveries in the Urals, were beginning to export oil to Western Europe for the first time since the 'twenties, in quantities carefully limited by the importing governments to avoid disrupting the markets. But Mattei was determined to launch an all-out price-war against his rivals in his home territory of Italy. He flew to Moscow to see Khrushchev, and from 1959 cheap Russian oil flowed into Italy, accounting at the peak for 16 percent of the market. It was a wonderful spectacle for the enemies of the seven, and the resulting price-war gave Mattei still greater popularity as the champion of cheap oil, at a time when the *autostrade* were spreading, and the little Fiats were multiplying across Italy: he even persuaded the Italian government to reduce the petrol taxes. It was a hideous embarrassment to the big companies, particularly Exxon and BP, Mattei's chief rivals in Italy. Their profits had already been pared down by the incursions of the independents: but they had to cut their prices further, all through Europe, and Mattei's price-war eventually pressed them to cut their 'posted prices' in the Middle East, thus precipitating — as we will see in the next chapter — the formation of OPEC.

Mattei, with the help of the Russians, still undercut them: Exxon, as the company most threatened, resolved to try to make peace. By 1962, through the mediation of George Ball, then Under-Secretary of State, they had begun to work out a truce, whereby Exxon would supply AGIP with cheap oil, and Mattei was invited to America, were he expected to be received by President Kennedy. But the peace-making proved unnecessary. In October 1962 Mattei's plane took off from an airfield in Sicily and crashed, with no survivors.

Mattei has ever since been a mythical figure to the Italian left, and rumours have circulated that Mattei's plane had been sabotaged by one of his many enemies. But such a drastic reprisal was not at that stage really called for. Mattei had already prepared to make peace, and his companies' financial base—as soon emerged after his death—could not support the fight much longer. In the meantime he had already done his damage to the sisters, which could not be undone. He had shown the producing countries that the cartel could be broken. He had forced the seven sisters to cut the profits of the producers, thus provoking the counter-cartel. And he had encouraged other consuming nations, too, to form their own national companies in rivalry with the majors.

The French company CFP, as we have seen, was already big enough to be regarded as a kind of eighth sister, and to be included in the Iranian Consortium. But its President, Victor de Metz, still smarted under his exclusion from Saudi Arabia, and French governments longed to be independent of the Americans. A great opportunity seemed to appear in Algeria, where CFP collaborated with an Algerian state company to look for oil in the Sahara, and found a huge field in 1956. The existence of oil in Algeria increased the determination to hold on to it despite the long Algerian War for independence, and the French suspected the American oil companies of encouraging their governments' support of the rebels. After peace in 1962 the French hoped that oil could be part of a special relationship with independent Algeria, but the Algerians were determined to control their own oil, and eventually they were to nationalise all their oilfields.

CFP were steadily expanding their sources and markets elsewhere, and their sign TOTAL was spreading through Europe and Africa. It was also competing with another French company, called ERAP, created in 1966 by merging two state-owned organisations, and selling its own gasoline under the name ELF: in the same year ERAP made an important new deal in Iran, agreeing to be simply a contractor for new exploration, without being a concessionaire. The French were thus, after the long years of resentment, beginning to play it both ways—acting like a sister, but also an aggressive outsider.

The Japanese, too, were determined to have a direct stake in the Middle East, and to escape from their dependence on the big companies, and in 1957 they agreed with the governments of Saudi Arabia and Kuwait to search for oil in the offshore waters of the Neutral Zone. They offered generous terms, giving 57 percent of the profits to the governments instead of the usual 50 percent. They formed a modest company with only about $10 million capital, which was tactfully called not the Japanese-Arabian oil company, but simply the Arabian Oil Company. In spite of the gloomy forecasts of its rivals, it was soon exporting oil from Arabia to Japan and making good profits, with some help from the Japanese government in ensuring its markets. By 1965 it was supplying 15 percent of Japan's oil. The remarkable partnership across half the world mitigated Japan's painful awareness that its whole economic recovery was based on oil from a remote region dominated by others. The Japanese presence in Arabia soon made itself felt with a rush of cars, cassettes and cameras.

These national companies and others—including the Belgian Petrofina, the German Gelsenberg, the Spanish Hispanoil—all began to complicate the control of the seven sisters and their dominance over the producing countries; they stepped up competition, and added to the glut.

THE GLUT

The glut seemed endless, and the bargain fuel still further transformed the Western way of life. The Europeans took to cars and motorways with the same speed and enthusiasm as Americans thirty years before; and they hardly needed pushing by the oil lobbies. Motorways, autobahnen, autostrade, forged through the last wildernesses of Europe—down to Calabria, through the Midi, up to Scotland. Cars poured out in millions, and the oil companies proclaimed themselves in clamorous advertising as the standard-bearers of the great new car-culture, promising virility and glamour as well as speed and comfort.

The filling-stations, fluttering with bunting, competed with free gifts of glasses, toys or paperbacks, with 'gasoline games', or with stamps, double stamps, triple stamps. As competition mounted, the advertising reached a crazy crescendo of claims and counter-claims. BP advertised 'Getaway People' driving at breakneck speed through the continent; Regent (alias Texaco) offered a provocative girl in a cowboy hat, pointing a phallic nozzle at the customer, and saying 'Get out of town, quick'. Shell had white-booted Supergirls, and a lottery to 'Make Money with Shell'. And Exxon promoted the most famous of all campaigns, the Tiger in Your Tank, with the genial beast leaping out from hoardings in every language—a symbol not only of Exxon but of a materialist kind of Western unity, too. As the cheap oil gushed out of Saudi Arabia, Iran and Libya, so the seven sisters escalated their zany war between brands of indistinguishable petrol. The green shield of BP, the yellow badge of Shell or the flying horse of Mobil popped up across Europe, while new signs proclaimed the intruders: the six-legged dog of AGIP, or the makeshift signs of Jet or Sprite, the torch of Amoco, the red letters of Total.

The roadside extravaganzas were part of a much deeper penetration of the oil companies into the European economy. Through the 'fifties and 'sixties, oil was taking over from coal as the chief source of energy in Europe, generating electricity, providing heating for homes and factories. The post-war European planners, led by Jean Monnet, had looked to coal as the essential basis both to rebuild Europe and to unite it; hence the importance given to the Coal and Steel Community. But between 1950 and 1965 the share of oil in providing energy for the six Common Market countries went up from 10 percent to 45 percent, while coal went down from 74 percent to 38 percent. The coal industry, unable to compete with the cheap oil, was rapidly run down, and miners regarded as expendable, while Europeans became increasingly dependent on fuel from outside their frontiers.

The Common Market countries appointed three wise men, headed by Louis Armand, to report on the energy situation in 1957: they gave a grave warning about Europe's excessive dependence on Middle East oil, and advocated massive efforts to develop nuclear energy through Euratom,[13] but the oil companies lobbied energetically against it, and Europe's dependence soon increased much further. With hindsight, it seems astonishing that the continent should have based its prosperity on such a perilous supply: but few experts warned against this dependence. One of them was E. F. Schumacher, then economic adviser to the National Coal Board in Britain, who gave a lecture in April 1961 predicting an oil crisis, and warning that in twenty years' time the demand for coal would be greater than ever. But in the prevailing infatuation with oil, he was ignored.[14]

Oil was much cheaper than it appeared to the ordinary motorist; for Western governments had long ago learnt that they could impose huge taxes on the stuff without appreciably affecting the consumption: in most of Europe in the 'sixties, the tax amounted to far more than the cost of the gasoline—a fact which only gradually dawned on the producer governments until they eventually realised that, whatever taxes they were levying, the consumer governments levied far more.

In the United States the gasoline was still cheaper, with a far lower tax than in Europe, and the basis of a more spectacular change of life-style—enabling Americans to escape still further from the cities, into the suburbs, and ex-urbs. Here too the oil consumers did not need much pushing, but the oil companies had done what they could to eliminate rival transport. In Los Angeles, for instance, the first sprawling suburbs had been built not round cars and freeways, but round an electric railroad system, whose relics can still be seen alongside the freeways; but in the late 'thirties General Motors collaborated with the local oil company, Socal (and also with Firestone), to buy up the railroad, and soon afterwards to close it down.[15] They did similar deals with oil companies in other big American cities; ensuring that the inhabitants would be dependent on road-transport alone. The freeways cut further into the country and through the hearts of the cities, while the railroads decayed.

Cheap oil and gas could transform not only the landscape but the climate; and the post-war years brought a boom in air-conditioning. Today Houston, the oil capital of Texas, is a kind of monument to what oil and gas could achieve, in defying geography. In the muggy Texan summer, a Houstonian can spend his whole day inside strange-shaped structures with unchangingly equable weather: in a shopping arcade with a skating rink; in a Hyatt hotel built round its own cool bit of sky; or in the Astrodome, a vast concrete mushroom where he and 50,000 other Texans can watch the Houston Oilers playing air-conditioned football. Since Captain Lucas sweltered under the harsh sun at Spindletop, the Texans had triumphantly won the battle against the elements.

The West was coming to regard cheap oil as being as natural as cheap water, and even while the price came down, there were still plenty of people to complain it was still too expensive. The oil companies were bitterly arguing with the producing governments about each decimal point on a barrel. How could they, in the midst of this free-for-all, argue that producers should actually get *more*?

In their relations with the producers they now had the worst of both worlds. They provided a common enemy for them to unite against, still carrying their pre-war reputation for conniving and fixing the markets. Yet they were now powerless to prevent the headlong decline in oil prices, which was not only threatening their profits, but leading the whole Western world into a fool's paradise.

NOTES

1. Interview with author, January 15, 1975.
2. *New York Times*, February 6, 1974.
3. Interview with author.
4. Interview with author.
5. According to one American definition, an independent is a company which depends on outside supplies for more than 70 per cent of its oil; so that their characteristic (as Paul Frankel has observed) is not independence, but dependence.
6. Edith Penrose: *The International Petroleum Industry*, p. 133.
7. Quoted in Engler: *The Politics of Oil*, (Phoenix Books) Chicago, p. 234.
8. John Bainbridge: *The Super Americans*, London, 1962, p. 90.
9. Multinational Hearings: 1974, Part 7, p. 139.
10. H. S. M. Burns in a paper to the API, November 17, 1955.
11. P. H. Frankel: *Mattei: Oil and Power Politics*, London, 1966, p. 40.
12. Frankel: p. 61.
13. *Un Objectif pour Euratom*, May 9, 1957.
14. E. F. Schumacher: *Small is Beautiful*, (Abacus) London, 1974, pp. 97–110.
15. See Statement of Bradford C. Snell to the Senate Anti-trust Subcommittee, February 26, 1974.

8

OPEC

We have formed a very exclusive club . . . Between us we control ninety percent of crude exports to world markets, and we are now united. We are making history.

Perez Alfonso, 1960

In July 1960, in the Rockefeller Center in New York, the new chief executive of Exxon, Monroe Rathbone, faced a decision which he knew would have world-wide implications. Rathbone was a man who had come up through the traditional pipeline; educated as a chemical engineer, he had worked his way to the top of Standard Oil of Louisiana. He made his adopted home in Baton Rouge, the great refinery centre, before he moved over to Exxon in New York. He was not a man of great international sensitivity; but he was regarded as one of the more far-sighted of Exxon's bosses, with a reputation for taking advice from others.

Rathbone's problem was the glut. It was bringing about price-cuts all over the world, and the Russians, as he saw it, were using oil to upset the economy of the West with dangerous results.[1] In Italy, Mattei had just made a deal with the Russians to buy crude oil at sixty cents below the Middle East price; in Japan, oil was being sold at huge discounts, made more damaging because they were publicly known. Worse still, the Russian oil was now invading India as had happened to Deterding's price-war in the 'twenties. The Indian government had told the three big refineries—each owned by one of the sisters—that they had been offered Russian oil cheaper than the oil from the parent sisters, forcing the companies to undercut their own posted prices. For Rathbone this was the most serious aspect: 'The tremendous amount of price discounting to third parties,' he told a reporter, 'is bad enough. If it now spreads to affiliates, the fat's in the fire.'[2]

Exxon and the other sisters were caught in a painful fork. On one prong they were compelled to reduce prices in the market-place. On the other they were still sticking to the same posted prices which were the basis for their fifty-fifty taxes to the producing countries. so that they were having to pay the countries much more than fifty percent. Yet the countries had come to plan their whole budgets on the confident expectation of these taxes. The basic flaw in the fifty-fifty agreements, which had looked so neat was now very evident. They were like plans to give factory-workers a shareholding in a company; fine when profits were booming, explosive when they were slumping.

The companies managed to create the impression that their profits were falling dangerously low, but in fact they were still able to pay for their vast expansion from the profits from Middle East oil. 'Thoughout the period,' wrote Professor Edith Penrose, 'the majors were able to finance from their own funds the construction of an enormous volume of transport facilities, refineries and distribution networks, and succeeded in maintaining their dominating position in an industry expanding at around eight percent per year.' But the sisters nevertheless felt threatened, because the independents were making their markets less assured: 'certain of the normal functions of prices, long rusty with disuse, began to reassert themselves.'[3]

They had already made one cut in the posted price in February 1959, of eighteen cents a barrel—a reduction which meant that the four major producing countries in the Middle East would receive about ten percent less in taxes, or $132 million dollars a year. But it was not altogether unexpected, for basically it restored the price to the level before the increases negotiated by the oil companies after the Suez Crisis two years earlier.[4]

Now Rathbone, surveying the price-cuts round the world, resolved that he must cut the posted price still further. It would be clearly explosive. There were already warnings from the producing countries that they could not tolerate a further cut and the New York magazine, *Petroleum Week*, having got wind of Rathbone's resolve, emphasised that no-one 'was anxious to face the uproar such a move would undoubtedly precipitate'. There was also heated opposition from some of the Exxon directors. Howard Page, who was concerned with the Middle East, and politically the most sophisticated, was emphatic about the danger: all hell, he said, would break loose. If the board insisted, they should at least share the loss of revenue with the governments. Page's view was shared by another director, William Stott, who was in charge of marketing, and who thought that Exxon should increase its marketing profits, to compensate for its losses in production. But Stott and

Page were over-ridden by Rathbone and the rest of the board.

The original plan was to cut the posted price everywhere, including Venezuela. But Exxon's relations with Venezuela had recently been stormy, and when the board instructed their man in Caracas, Arthur Proudfit, to enforce the cut, he threatened to resign, believing that it would wreck Exxon's position in Venezuela, so the board climbed down. On August 8, 1960, Exxon made its announcement; that the posted prices in the Middle East would be cut by an average of ten cents per barrel. This historic decision, which so drastically diminished the income of the chief Middle East countries, was taken inside the boardroom of a private corporation.

The reaction of many Middle East old hands was one of horror—particularly within BP, with their huge stake in the region. Harold Snow, the vice-chairman, the silent mathematician who had patiently negotiated the Iranian agreement six years before, was observed by a colleague to be in tears when he heard the news. And the chairman, Maurice Bridgeman, took the unusual step of making a public statement deploring the price-cut: BP, it said, had hoped that such action could be deferred, and heard the news with regret. Shell believed that if there were to be a price cut it should also be made in Venezuela, and they nearly went it alone there; but they were dissuaded at the last minute.

The sisters were all so interdependent that they all had to follow Exxon. A few days later BP cut its price by $4\frac{1}{2}$ cents and the other companies hovered between different prices until they were all down, and all level. The companies were solidly confronting the Middle East.

All hell *did* break loose.

There had been earlier attempts to bring together the oil exporting countries. The Venezuelans, as we have seen,[5] influenced by the shrewd Perez Alfonso, had for long been determined to safeguard their own conservation policy by spreading it abroad. With their very limited reserves, their long history of independence, and their bitter resentment of the United States, they were much more militant than the Arabs, and their proselytising had helped to establish the principle of 'fifty-fifty' in the Middle East. The Arabs had a growing, if vague, realisation of the importance of collaborating over oil, ever since the Arab League was formed in 1945 with oil in its terms of reference.

President Nasser of Egypt, in his own book *The Philosophy of a Revolution* first published in 1954, described petroleum as one of the three chief components of Arab power. Like so many Arab radicals, he had got his ammunition from America. He had then just been reading a treatise on petroleum published by the University of Chicago, which revealed to him that it cost only ten cents to extract a barrel of oil from Arab countries; and that the average output of a single Arab oilwell was 4,000 barrels a day, compared to eleven barrels in America. 'Petroleum', Nasser told his Arab brothers, 'is the vital nerve of civilisation, without which all its means cannot possible exist.' But the Arabs with the most oil, led by the Saudis, were not keen on sharing their revenues with Egypt who had almost none, and it was not until 1958 that there were the beginnings of some unity. The Iraqis, who were soon negotiating with the IPC, felt the need for support from other Arab countries. The new oil adviser in Saudi Arabia, Abdullah Tariki, was a radical who was determined to unite the producers. Tariki was one of a new generation of Arabs who had seen the oil industry from the inside; he had been educated at the University of Texas, had been a trainee in Texaco—a bitter ordeal—and was married for a time to an American girl. He was left with a detailed knowledge of oil and of the anti-trust arguments inside America: once again, America was educating its adversaries.

The first price-reduction in 1959 had provided the first spur to unity. Two months later the first Arab Petroleum Congress was convened in Cairo, including observers from Iran and Venezuela, and recommended that there should be no reduction in the posted price without consultation with the governments. And there were private talks between Tariki and Perez Alfonso, which led to a secret gentleman's agreement which, according to Alfonso, 'constituted the first seed of the creation of OPEC'. But the oil company men at the congress were not too worried, and Iraq, the most militant producer, was too angry with Egypt to attend.

But the second reduction, led by Exxon, transformed the whole atmosphere in the Middle East. In Iraq, it came just when the oil companies were deadlocked in their negotiations, and the Iraqis suspected it was aimed at bringing pressure upon them. They were determined to fight back, and on September 9, a month after the announcement, they convened a meeting at Baghdad of five countries who were responsible for eighty percent of the world's exports of oil: Saudi Arabia, Iran, Iraq, Kuwait and Venezuela. The key alliance was between Saudi Arabia and Venezuela, and the meeting was dominated by Tariki and Perez Alfonso. The two men already trusted each other: Tariki had been much influenced by Alfonso's arguments about conservation, and the fact that Venezuela had been excluded from the posted price cuts enabled Alfonso to show the advantages of being tough, and made him more determined to show unity with the Arabs.

The Shah, too, was furious about the price-cuts of which he was given no warning: 'Even if the action was basically sound,' he said afterwards, 'it could not be acceptable to us as long as it was taken without our consent.'[6] In his indignation he was now prepared to join up with the revolutionaries of Iraq, and the Shah's support was very important to the new movement. But he still kept aloof from the Arab members, continuing to supply oil to Israel,[7] and he was not at all confident of the power of the producers: 'I must admit we were just walking in the mist,' he explained to me fifteen years later: 'not in the dark, but it was a little misty. There was still that complex of the big powers, and the mystical power and all the magic behind the name of all these big countries.'

The new price-cut was most serious for Iran, which had far the biggest population, but it also came at a bad time for Saudi Arabia which was just embarking on a massive expenditure on social services. Some other Arabs did not mind the thought of the Saudis having to cut down: thus the *Egyptian Economic and Political Review*, which was close to the Arab League, commented with some gratification that the loss of revenue 'should mean fewer Cadillacs in Arabia'.[8] But among the oil producers the price-cut generated a surge of unity, between the conservative kingdoms of Saudi Arabia, and the anti-monarchists of Iraq; and between the Arabs and two states, Iran and Venezuela, right outside the Arab world.

THE BIRTH OF OPEC

The five nations met in Baghdad in a mood of excited confidence: 'It was quite clear from the start', wrote the ebullient owner-editor of the *Middle East Economic Survey*, Fuad Itayim, 'that the price cuts might precipitate the establishment of what some delegate chose to call "a cartel to confront the cartel". It had precisely that effect!' They decided on the foundation of the Organisation of Petroleum Exporting Countries, or OPEC. The first resolution made clear that their chief enemy was the oil companies, and they stated:

that members can no longer remain indifferent to the attitude heretofore adopted by the oil companies in effecting price modifications; that members shall demand that oil companies maintain their prices steady and free from all unnecessary fluctuation; that members shall endeavour, by all means available to them, to restore present prices to the levels prevailing before the reductions . . .

The preamble expounded the common predicament of all the oil exporters, with unexceptionable logic: that they all depended on oil income to finance their development, and balance their budgets; that oil was a wasting asset, which must be replaced by other assets; and that fluctuations in the price of oil dislocated not only their own economies, but those of all consuming nations.

Would OPEC, or some such body, have formed itself at about this time, without the urgent prodding from the oil companies? Certainly the tide was already flowing towards militancy. The tide which moved the Mexicans to nationalisation, which gave the Venezuelans their toughness, and which swept Mossadeq into power in Iran was now carrying the Arabs to the awareness of their potential, and these countries would have found common cause before long. But the critical point about Exxon's decision was that it was unilateral, with no attempt at consultation; and it compelled the countries to respond in the same fashion. Confronted by the unity of the seven, they had to oppose with the same kind of unity. As one Kuwaiti insisted to me, 'OPEC couldn't have happened without the oil cartel. We just took a leaf from the oil companies' book. The victim had learnt the lesson.'

The founders of OPEC owed much to the oil companies, and to their critics in the West. The new generation of young Arab technocrats like Tariki knew the history of conservation in Texas, and they had followed carefully each attack on the 'oil cartel'. The facts from the Federal Trade Commission's report in 1952 were repeated through OPEC speeches. Like the anti-colonialists in the British Empire, they took their weapons from their masters, and the awakening could be interpreted as a reflection of American democracy as much as of Arab nationalism. The concept of the cartel provided an ideal common enemy; and the oil companies were very conscious of being isolated and unloved. As one director of Shell recalled: 'OPEC could count on some support from the consuming countries in putting forward their notion of a cartel. When the industry was attacked, it was without friends.'

OPEC began with a flourish. A month after its foundation, the

Arab Petroleum Congress held another meeting in Beirut, attended by oil-company men. Tariki took the opportunity to lash out at them, accusing them of rigging their profits to deprive the producing countries of more than two billion dollars over the previous seven years. He accused Aramco of having concealed from the Saudi Arabians that they received a discount on their oil; so that the fifty-fifty agreement was really 32–68. The oil companies protested, but would not reveal the true figures: the extent of their Middle East profits, buried in their complex accounting, remains hidden to this day. Tariki had made his point to the other delegates, that the companies were concealing the facts, and the common indignation gave a new fillip to OPEC.

The new organisation was duly set up in Geneva, with an Iranian Secretary-General, Fuad Rouhani. He was an urbane and very moderate man, an international lawyer and a talented musician and he was determined (as he recalled to me) to keep OPEC out of both politics and religion: he would tour the offices to make sure that officials stuck to oil and economics. But he soon realised that the oil companies were determined to pretend that OPEC simply did not exist; and insisted on negotiating separately with each country. So he turned up at each negotiation, 'wearing a different hat'. While the companies insisted that OPEC was a would-be cartel, Rouhani observed that the companies were clearly in cahoots, co-ordinating their terms with each country.

From the beginning, OPEC achieved one important aim: it prevented further reductions in the posted price, even though competition was becoming still more intense. As Nadim Pachachi, a later Secretary-General, described it, OPEC froze the 'managerial prerogative' of unilateral fixing of prices. The companies tried to get individual countries to give tax rebates in return for extra production: but they stood firm and resisted the temptation.[9] But it did not restore the posted prices to the previous levels, which was another of its aims. And, more serious, its members could not agree about how to fix prices, or to restrict production. At the second conference, in Caracas in January 1961, OPEC resolved to make a detailed study to arrive at a 'just pricing formula'. But each of the members desperately wanted a bigger share of the market, and the notion of conservation, however much preached by the Venezuelans, had little appeal in the free-for-all. The countries were back in the predicament of the oil-drillers in Pennsylvania a century before: divided and ruled by the men who controlled the market.

And the Russians, who could have helped the Arabs to cut back the glut, were still spoiling the market. At one point they actually showed interest, through Rouhani, in becoming members of OPEC, but they took it no further. Instead they explained to successive meetings of the Arab Petroleum Congress that they wanted to get back their position as major oil exporters, and they would not support artificially high prices. For the time the Russians were the chief bogey, both for the Arabs and the companies, and the Americans became increasingly worried that the Russian 'dumping' of oil was an attempt to sabotage the West. In November 1962 the National Petroleum Council in Washington produced a long report analysing the dangers, which reckoned that Soviet exports had cut the income of producing countries by $490 million in five years. However the Russian threat soon afterwards faded away, as the Russians needed more oil for themselves and their satellites; by the late 'sixties they were actually importing oil from the Middle East.[10]

The first surge of OPEC unity did not last long. The Iraqis, preoccupied with their own problems, did not turn up at the meetings. The radical Arabs resented the moderate Rouhani: he was eventually succeeded by an Iraqi, Abdul Rahman Bazzaz, who encouraged talk of both religion *and* politics. Tariki fell into disfavour and was replaced as oil minister in March 1962 by a young lawyer who seemed a much more pro-Western influence. Sheikh Zaki Yamani, then only thirty-two, was the son of a judge in Mecca, who had studied law first at Cairo, then at New York University, then at Harvard. He had come back to Saudi Arabia, sophisticated and westernised, to start a private law practice.

His appointment as oil minister and as director of Aramco was very welcome to the four companies in Aramco: he became friendly with the directors, and loved visiting New York. And the State Department, as well as Aramco, had taken pains to cultivate the Saudi royal family. King Saud, who had succeeded his father Ibn Saud in 1953, had been an increasing embarrassment to the diplomats: but by 1962 he had been ousted by his brother King Feisal, whose hooked nose and sinister expression had so struck Dean Acheson sixteen years earlier. King Feisal was still implacable on the question of Israel; but he was also deeply conservative and very wary of the Arab radicals, particularly after Nasser was threatening the Saudis through the war in Yemen. The King looked to Washington as an anti-Communist ally, to maintain the most special of special relationships. Moreover King Feisal, though his revenues were increasing, could always spend more, and had no wish to hold back production to help his Arab rivals.

OPEC was still totally failing to achieve any kind of effective pro-rationing of oil production between its members, of the kind that had maintained prices in Texas for the past thirty years. Saudi Arabia was becoming the Texas of the Middle East: if they would not restrict production, no-one else would. But they showed no interest in restriction, and the problem was now harder than in 1960, for new oil-producers were starting up all the time, including Nigeria, which was right outside OPEC. Iran and Iraq were proposing 'programming' production according to population—both having large populations—but others wanted programming according to need. At the ninth OPEC conference, in Tripoli in July 1965, members set limits for each country's increase on an experimental basis: Saudi Arabia and Iraq would go up 9 percent in the next year, Iran 14 percent, Kuwait 5 percent, Libya 33 percent. But Saudi Arabia was not at the meeting, having had a row with Iraq about an apology, and the Saudis vetoed the decision: their increase for the year was in fact 18 percent. 'The argument was really nothing to do with programming as such,' said Yamani afterwards, 'it was about resources and economic factors.'[11] By 1967 OPEC had given up the attempt.

The companies were not now seriously worried by the threat of OPEC. They continued to deal separately with each government, and to play them against each other; while the governments remained in some awe of the power of the companies. But OPEC did have one significant achievement. In June 1962 they resolved to establish a uniform rate for royalties in each country, which would not be deductible from the income tax paid to them. The companies resisted stubbornly, claiming at one point that it would reduce their profits by a quarter, and the argument dragged on for over two years. Eventually in Geneva in December 1964 they reluctantly agreed, thus enabling the producers to increase their revenue at a time when the price was going down. To the oil companies it seemed a generous concession, involving an extra four cents a barrel, and to OPEC it seemed a great victory. But they had not touched the central question of the price of oil, for they were quite unable to control the flow of it.

One man who did take OPEC seriously was the veteran lawyer-administrator, John Jay McCloy, who was then regarded as a kind of chairman of the American 'Establishment'[12] and was to become the key figure in oil diplomacy. McCloy, then sixty-six, had moved effortlessly through the revolving doors of government and business. His origins were modest; he was born, as he liked to recall, on the wrong side of the tracks in Philadelphia, and worked his way up through Harvard Law School. Patient, philosophical and humorous, he was a natural mediator, and before long he was a top lawyer in Wall Street, and a confidant of the Rockefellers. During the war he was Assistant Secretary of War, and he had been High Commissioner for Germany, President of the World Bank, and Chairman of the Chase Manhattan.

When President Kennedy took office in January 1961 McCloy advised him on such questions as arms control, security, defence and Germany. At the same time he was practising as a very highly-paid lawyer, in the prestigious firm of Millbank, Tweed, Hadley and McCloy, and from that office, it later transpired, he represented the anti-trust interests of all seven of the seven sisters.[13] 'My job', as he described it to me later, 'was to keep 'em out of jail.' In the midst of a world of conflicting and disconnected interests, McCloy appeared as part of that discreet 'supra-government' which remains while Presidents come and go, and it was natural that Kennedy should turn to him for advice on oil, too.

Kennedy was concerned about a Middle East confrontation with the Russians after his talk with Krushchev in Vienna in June 1961. He talked to McCloy who then warned Kennedy about the possible consequences of OPEC: if OPEC were to succeed in joint action, he said, it might be necessary for the oil companies on their side to be given authority for collective bargaining. Kennedy 'right then and there' arranged for McCloy to see his brother Robert, the Attorney-General, to whom McCloy repeated the warning. Robert Kennedy assured him that, if and when the companies contemplated joint action, he would be glad to discuss the possibility. McCloy thereafter made it his business to call on each successive Attorney-General, to repeat the warning: first to Katzenbach, then to Clark, then to Mitchell; though it was a decade before the expected eventuality arose.[14] The principle was established: that for the sake of security of oil supplies and for reasons of foreign policy, the anti-trust laws would be waived again.

THE BALLOON

While OPEC was failing to control production, it was the sisters who were trying to reconcile the demands of the countries, and in effect determining their future growth rates. Each country was desperate to *push up* production to secure a higher revenue without worrying too much about the world low price; the companies were trying to *hold back* the production, to prevent the price going lower—the precise opposite of the apparent situation a decade later.

It was clear that in the midst of the glut the companies were agreeing between themselves to hold back production. However the producing governments could not discover exactly how they were taking their decisions, and their arrangments were carefully shrouded. What was evident was that, since the formation of the Iranian Consortium (which included all seven or eight) all the sisters were interlinked in joint ventures, which made the task much more manageable. At the centre of the ring was Exxon, who was a partner in Aramco, and in the Iranian Consortium, and in the Iraq Petroleum Company: and the ringmaster was their Middle East negotiator, Howard Page, their director for twenty years, who had already shown his ingenuity in Iran in 1954. Patient, conciliatory and calculating, it was Page who masterminded the strategy to restrain the demands, and to divide and rule the potentates.

Pages' provenance was unlikely for diplomacy: he was a Californian brought up at the refinery at Oleum, near San Francisco, which his father ran for Union Oil; and to young Howard the acrid smell of a refinery had always been the smell of home. He rose up, like most senior oilmen, through the pipeline of chemical engineering, and designed refineries in Texas. Then he joined Exxon, who posted him to supervise refineries in Europe, where he encountered a quite different world. He lived in Paris, married an English girl working for Exxon, and learnt to speak French. Dealing with Continentals he became aware of quite other views: 'I began to think that perhaps they couldn't all be wrong except me.' After the war, when he worked in Washington, he began to work for Exxon in the Middle East; coming fresh to it, he did not have the fixed assumptions of the old timers, or any emotional involvement with 'the sanctity of contracts': he was essentially a pragmatist. He had learnt his cardinal rule in negotiating: always prepare beforehand a range of alternative solutions so that, if one proposal breaks down, you can be immediately ready with another: sometimes the most unlikely alternative was the most acceptable. With this experience behind him, Page came to Teheran, a quiet, unassuming man, with an easy smile and sharp eyes; the look of a genial lizard.

Page saw the market like a bulging balloon; 'push it in one place, it comes out in another ... If we acceded to all those demands, all of us, we would get it in the neck.' The balloon was pressed harder with each new discovery. New sources of oil were opening up along the Gulf, in Qatar, in Dubai, in Oman, and most spectacularly in the small Sheikhdom of Abu Dhabi, which by 1970 was producing a million barrels a day. And the most irresistible new opportunity was in Libya, where Exxon was the leader. But the more they took from the new sources, the less they could take from Iran or Saudi Arabia. Although Exxon was spending huge sums on new exploration, the last thing they wanted was new production. Howard Page was once told by one of the Exxon geologists, who had just come back from Oman; 'I'm sure there's a ten billion oilfield there.' Page replied, 'Well then, I'm absolutely sure that we don't want to go into it, and that settles it. I might put some money in it if I was sure that we weren't going to get some oil, but not if we *are* going to get oil, because we are liable to lose the Aramco concessions.'[15]

Any country that made too many difficulties for the oil companies provided a useful excuse to cut back; as Howard Page put it: 'Sometimes they made it easy to cut down by breaking an agreement, as in Iraq; then we could tell 'em to go to Hell.'[16] Iraq was held out as a new warning to the others, like Iran under Mossadeq, to 'co-operate, or else'.

But in fact Iraq had had good reason to be sceptical of the oil companies' good faith, ever since the Iraq Petroleum Company had begun operations in 1928. The company had refused to grant the Iraqis 20 percent participation, and had repeatedly delayed new production,[17] against the stipulations of the original San Remo agreement. The delaying tactics of the IPC had been criticised in the Federal Trade Commission's report of 1952, but this portion had been excised for reasons of 'national security', and was not published until 1974.[18]

The resentment of Iraq against the companies became much fiercer after the revolution of 1958. Iraq helped to radicalise the other Arabs, and to bring about OPEC, but in the meantime her

THE DANCE OF THE SISTERS

OWNERSHIP LINKS BETWEEN THE MAJOR INTERNATIONAL OIL COMPANIES INCLUDING COMPAGNIE FRANCAISE DES PÉTROLES (CFP) AND THE MAJOR CRUDE OIL PRODUCING COMPANIES IN THE MIDDLE EAST

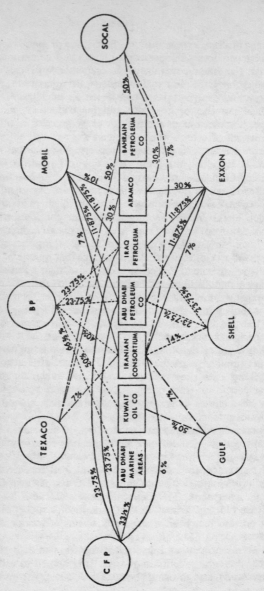

Source:- Multinational Hearings Part 5

oil production lagged behind—even though her reserves were bigger than Iran's. With the past history of delays and conceal-ments, the Iraqis promulgated in 1961 the contentious Law 80, which proposed to nationalise most of the concession. It began a long period of uncertainty, and the confusion increased after the assassination in 1963 of Qasim, and his replacement by a success-sion of dictators. The oil companies, the partners in IPC, all now had large sources of oil elsewhere, and most were quite glad to restrict production and investment, and even to disinvest from Iraq. If Iraq had produced more, Page's task would have been harder; as it was, when he was asked 'can you swallow this amount of oil?' he replied, 'Of course, with Iraq down.'[19]

The two main oil-producers were now Iran and Saudi Arabia, and in spite of their coming-together in OPEC they were old rivals with quite different motivations. Iran had a population of thirty million, relatively short-lived oil reserves, and a Shah with an impatient ambition to modernise, industrialise and militarise his nation. Since his 'White Revolution' of 1963 which had begun to redistribute land, the Shah had for the first time felt sure of his throne and the support of his people, and he was more determined to show his independence from the West.

Saudi Arabia was thought to have only around six million people (there had been no proper census) with vast oil reserves, and with sporadic ideas about what to do with the money. Each country wanted to be master in the Persian Gulf—or the Arabian Gulf. Exxon and the other sisters tried to hold the ring between them, to prevent too much oil reaching the market. The balancing-act was made much easier by the fact that the four sisters in Saudi Arabia were also partners in the Iranian Consortium: with Exxon as the chief go-between.

Ever since 1954, when Iran began exporting oil again after the Mossadeq crisis, the balance had been precarious. Howard Page had persuaded the old King, Ibn Saud, with some difficulty to restrict his expansion to make room for Iran, to prevent it from going Communist.[20] The Shah first insisted on a fixed rate of increase of eight percent a year, but Page eventually dissuaded him on the grounds that anything above that fixed rate would be demanded by the Saudis, and gave the Shah instead a promise that the Consortium's growth would keep pace with the Middle East.[21] In fact, with the prevailing oil-boom, the Iranians soon increased production by much more than eight percent. But as

the Shah found his confidence and learnt the game, he became
more ambitious; and he realised that Iran had not regained its
position before the Mossadeq crisis of being the biggest oil-
producer in the Middle East.

But it was a long time before the Shah learnt the secret system
of how the companies actually agreed to restrict production.
Both Iran and Saudi Arabia had secret 'offtake agreements' by
which a partner was penalised if he took more oil than a care-
fully-calculated mean of all the partners' total demands: so that
the biggest companies could set a norm for production, above
which the others would have to pay a higher price. The big companies could justify this commercially on the grounds that each
had contributed its share to the capital costs of development, and
it was unfair for any company to take more than its share without
paying extra.[22] But the mechanism was really a device for restricting production, and as such it was politically explosive.

In the first place the companies that desperately needed more
oil were exasperated by the companies that had too much oil—
notably Exxon, Socal and Texaco. In Aramco it was those three,
the 'thirty percenters', who wanted to hold back production, while
Mobil, who had only a ten percent share, consistently wanted
more oil. Mobil even threatened to do a deal with BP to take oil
from Kuwait unless the agreement was modified. In Iran, it was
again Exxon, Socal and Texaco, together with Gulf ('the four
bears') who wanted to hold back production: while the French
company CFP and the American independents were continually
trying to increase it. The big companies tried to close their ranks,
while the newcomers tried to break in, fighting both the giants and
each other: 'These larger companies showed us no mercy,' said
E. L. Shafer of Conoco, 'but I also did not see any mercy from
any of my brother independent competitors.'[23]

But in the second place these restrictive agreements were ammunition to the producing governments, for they implied that the
economic growth of their countries was being secretly dictated by
the boardrooms of private companies: and in the growing mood
of nationalism and self-awareness this was increasingly unacceptable.

The Saudis were becoming more worried by the Iranian competition, and King Feisal could not make full use of his oil revenues
with his long-term plans for economic development. His oil
minister, Yamani, suspected that Howard Page was basically pro-

Iranian, having negotiated the Iranian settlement, and always suspected the Shah of getting a better deal. Page had to persuade both sides that he was not playing favourites, and even resorted at one point to the different Iranian calendar in order to avoid straight comparisons. 'It wasn't really a system,' said Page's deputy and later successor as Middle East director, George Piercy, 'it was really a game of catch-as-catch-can.'[24]

But the Shah was the more demanding. By 1966, he had become impatient of restraints, particularly in the light of his ambitious fourth five-year plan and his growing military expenditure. He inspired a press campaign attacking the companies and threatening to deprive them of part of the concession, or to sell oil to the East. The Shah was convinced that the Iranian agreement was more restrictive than Saudi Arabia's, and he complained to BP, who complained to the British government, who in turn complained to the State Department. In October 1966, the five American sisters went to Washington to seek help from the State Department, who promised that the American Ambassador would do his best to keep the lid on the Shah's demands. But the companies made clear that they did not want the government involved in the substance of this problem.

The Shah was still unsatisfied and the situation was precarious. On November 16 a very specific warning reached Eugene Rostow, the Under-Secretary for Political Affairs in the State Department, from Walter Levy, the international oil consultant in New York. Levy told Rostow about the Iranian secret offtake agreement, revealing that it carried much higher penalties for 'overlifting' than the Saudi Arabian agreement; if negotiations broke down, he warned, 'the very fact that a restricted secret agreement exists would be political dynamite in the hands of the Iranians.' But a State Department memo stressed that the agreement was 'a highly sensitive, inter-company commercial policy matter' and it was 'desirable for the U.S. government to limit its involvement'.

Levy's apprehensions were shared by others, including the British Ambassador in Iran, Sir Denis Wright, and they soon proved well-justified. The French partners in the Consortium, CFP—infuriated by the restrictions on their own production—soon leaked (their partners alleged)[25] the secret partners' agreement including the controversial clause, to the Shah, and the fat was in the fire. By September 1967 Howard Page, on behalf of the big companies, was forced to modify the agreement to allow a 'half-way price': which halved the extra cost to the partners of oil lifted beyond the mean. This still did not satisfy either the Shah or the smaller companies—who both had the same interest in increasing production—and soon afterwards the Shah learnt that Aramco had settled on a 'quarter-way' agreement which still further reduced the cost of extra oil: the other partners suspected Mobil, now desperate for more oil, of leaking this one.

The Shah complained to the State Department, and the oil expert in the American Embassy, Robert Dowell, added his own misgivings in a revealing letter to the State Department's oil expert, Jim Akins. Dowell reckoned that the American sisters, particularly Texaco, were using the agreement to do down their competitors, and complained that 'the net practical effect smacks of restraint of trade'.

Akins, in his reply, was shocked by Dowell's use of that phrase, and his reply was equally revealing about sensitivity to criticism of the oil companies.

> We were surprised to see the accusation put into a telegram which was given fairly wide distribution in the government. But fortunately the strong anti-oil company voices in Treasury and Commerce do seem not to have picked up the matter. They couldn't get very far, in any case, given the fact that the American companies were given specific exemption from anti-trust legislation when they entered the Consortium.

Akins could not agree with the Shah's complaints, and suggested that if the Iranians were too aggressive, they would share the fate of Iraq. The companies reluctantly agreed to a compromise. The Consortium would relinquish a quarter of its concession, and give up some oil to the National Iranian Oil Company; who in turn agreed that, to avoid upsetting the Western markets, it would only sell its oil in exchange for goods behind the Iron Curtain.

Was this balancing-act simply a discreet extension of the old cartel of the seven? John Blair, the irrepressible economist who had been responsible for the famous Federal Trade Commission report of 1952, later produced his own analysis of the new system. He maintained that the period from the late 'twenties until the late 'forties was the 'cartel era' in international oil, while the postwar era was the period of 'oligopolistic interdependence'. The seven majors, he insisted, without necessarily direct collusion, shared common assumptions about the rate at which the industry *should* expand. The seven companies had achieved an amazingly stable increase in output between 1950 and 1972, averaging at 9.55 percent a year. In spite of sudden drops from individual countries, like Iran in 1951 or Iraq in 1957, or sudden new sources, like Libya or Nigeria, the majors were able to 'orchestrate' the countries into 'a smooth and uninterrupted upward trend in overall supply.'[26] And the majors instinctively limited their expansion and underestimated demand, knowing that their rivals were doing likewise, according to the pattern of 'imperfect competition'.

Certainly this control system, as Dowell complained, 'smacked of restraint of trade'. The joint ventures between the seven sisters in each country provided a network of understanding and gave each company an interest in not damaging the others. But unlike the pre-war cartel the system was visibly unsuccessful in keeping up prices to artificial levels; and by the late 'sixties it was hard to see (as Howard Page insisted) how prices could have fallen much lower. If more independents had moved into the Middle East, with a freer hand, they could have done more quick deals to sell surplus oil cheaply, in countries like Libya. But they would also be much more vulnerable to pressures and squeezes from the producing countries as was to happen in Libya (see Chapter 10). It is quite possible, as many oilmen argue, that the rule of the seven, as opposed to seventy-seven, could keep prices down, by holding a firm front against the claims of the producers. It may have been a cartel, but it was a cartel on the side of the consumers. Its principal purpose, in other words, was to screw the producers.

Certainly to the producing governments it was an increasingly sinister cartel, for it was keeping prices low at a time when most other commodities were rapidly climbing. They began to realise that, faced with a more fragmented industry, the producing countries would be much stronger. And if they wanted an example to prove it, they only had to look across Africa, to Libya.

THE SIX-DAY WAR

In the meantime the growing tensions between Israel and the Arabs were threatening to short-circuit the two opposite elements of American foreign policy to produce an explosion. On June 4, 1967 Israel invaded Egypt. President Nasser immediately claimed that Israel was supported by Britain and the United States, and the foreign ministers of the Arab states gathered in Baghdad. King Feisal of Saudi Arabia, who had only three weeks before talked in London about his fears of Nasser's aggression, was now committed to support his Arab brother in time of need. Following the lead of Iraq, the Arab states agreed to shut down the oil-wells and to boycott the Western imperialists. The war, it seemed, had given the Arabs the crucial incentive to unity.

But it was very short-lived. The oil producers quickly realised that they were damaging themselves more than anyone else, for two key members of OPEC, Iran and Venezuela, had no intention of joining the boycott and were soon benefitting from the resulting shortage. King Feisal was faced with an imminent financial crisis, and on the advice of Yamani he quickly limited the boycott to the two countries that were regarded as the aggressors, Britain and the United States (neither of whom anyway then took much oil from Saudi Arabia). By the end of the month, after the Saudis had lost $30 million in revenue, Aramco was allowed to resume normal exports; and the other Arab states followed.

The end of the Six-Day War left the Arab members of OPEC much worse off than before. Iran had taken advantage of the boycott to push up her exports to Britain and West Germany; Venezuela, too, had moved further into Europe; and even Libya was able to export more to West Germany. The United States, which was supposed to suffer most from the boycott, had hardly been touched. And the closing of the Suez Canal which resulted from the war had not seriously damaged the West, for the new giant tankers anyway went round the Cape. Sheikh Yamani admitted that the use of the oil weapon had been a fiasco: 'If we do not use it properly, he said,' 'we are behaving like someone who fires a bullet into the air, missing the enemy and allowing it to rebound on himself'. At an Arab Summit in Khartoum, two months after the war, the oil-producing countries had to face up to their failure, and each agreed to follow its own policy for the export of oil.

Some of the Arabs also decided to form their own club parallel with OPEC, called OAPEC, which first met in Beirut in September 1968. It seemed at the time a retreat from political action. The first members were Saudi Arabia, Kuwait and Libya (still under King Idris) who had all been damaged by the failed boycott, and excluded Egypt and Syria, while Iraq refused to join. The first

Secretary-General was Sheikh Yamani, and he and the Saudis insisted that the organisation must keep out of politics. By 1970 after the revolution in Libya, OAPEC had taken on a more radical character, with new members including Algeria, and eventually the Saudis were compelled to agree to let in Iraq. Bold projects were approved, including an Arab tanker fleet, an Arab dry dock in the Gulf and an Arab service company. But OAPEC was still not taken very seriously by the oil companies, for it was a forum for all the squabbles between its disparate members, and it seemed to be undermining OPEC rather than enforcing it.

The Shah was determined to extract the maximum reward for not having joined the Arab boycott. By November 1967 he was insisting to the State Department that he must have an increase in production of twenty percent a year, compared to the past average of twelve percent. This would be partly as a reward for running 'grave political risks', but also (as the diplomats privately revealed) because he had miscalculated the oil income for his fourth five-year plan—an error which he could not publicly admit. He now wanted his oil income to be linked to the cost of the plan. The State Department replied that the companies could not allocate production in terms of reward or need.

But in March 1968 Eugene Rostow nevertheless called the five American sisters to the State Department, headed by Ken Jamieson and George Piercy from Exxon, together with two independents, Gendron of Atlantic and Shafer of Conoco. Rostow explained that the Middle East situation had deteriorated and that Iran might join a future Arab embargo if the companies refused to increase production. The sisters replied that Iran must be given equal treatment with Saudi Arabia, and the independents complained that they were still not given enough incentive to lift extra oil. The Shah insisted on more revenue and as a desperate expedient the companies shifted their annual payments from the Gregorian calendar to the Persian calendar to buy three months' time.

As for OPEC, the whole basis of its unanimity now seemed to be in ruins. The Shah was trying to make gains at the expense of the Arabs, while the Arabs, after the fiasco of the war, were again disunited. Nasser and Egypt were discredited in the eyes of the Saudis, and the Aramco partners were again confident that they could safely depend on Saudi Arabia for increased supplies. King Feisal would not again ally himself with the militant states to the north. 'Exxon seemed certain that the Arabs could never get together', recalled a former Exxon executive, "their image of the Arabs was taken from the film of *Lawrence of Arabia.*' And in Washington, too, the sensational victory of Israel had encouraged disdain for the Arabs, and the conviction that oil and Israel could still be kept in separate compartments.

OIL FROM THE NORTH

OPEC was in disarray, but Exxon, having helped to provoke the formation of OPEC, was now taking precautions lest it might one day gain real power and control. After long arguments within the board, in the early 'sixties, Monroe Rathbone, the chief executive, pushed forward a massive programme for exploration. It was to cost $700 million in the three years from 1964, and it concentrated on territories outside OPEC. (There is some doubt among Exxon executives as to how far this exploration was caused by political prescience, or how far it was induced by the need to plough back the very high profits.) The programme was eventually successful, and was to put Exxon further ahead of its rivals. By the early 'seventies they were getting oil from S.E. Australia, and opening up new fields in the Mackenzie River delta in Canada; and most important of all, they had found vast reserves in Alaska and the North Sea.

Exxon, later followed by the others, soon faced more intimate political problems as the price of their success. These discoveries closer to home did not much diminish the West's dependence on OPEC, but they brought the sisters face to face with their home governments, and raised more acutely the political question of how to control them.

The Alaska discovery was not in fact pioneered by Exxon, but by one of the most enterprising of the independents, Atlantic (soon to become Arco). The company had recently been revived by a daring financier, Robert Anderson, who realising its desperate shortage of crude oil, gambled on exploration in the Arctic. In June 1968 Arco struck oil in Prudhoe Bay—a reserve which was eventually to prove as important as Texas. But Arco ran short of money, and the classic story of Standard Oil repeated itself. They were forced to bring in Exxon, who became the dominant partner.

BP, in the meantime, had been looking for oil in Alaska for nine years. Like Exxon, they wished to reinsure themselves against their dependence on the Middle East. They were desperate to have a stake in the North American oil business and having been unable to buy an American company they had gambled on finding indigenous oil. Nine months after the Arco find they too struck oil in Alaska—a historic breakthrough for a British intruder. The whole of this vast new reserve was controlled by three companies, with BP owning fifty-two percent. But BP had the problem of finding outlets for the coming flood of oil. After some anti-trust difficulties they acquired part of the Sinclair Oil Company, and soon afterwards they arranged a complicated deal progressively to take over Standard Oil of Ohio, another of the original Rockefeller companies, with a large network of filling-stations in its home state.

This in turn attracted the attention of Washington, and of the ardent anti-trust chief Richard McLaren, then in the first flush of his crusade against mergers, who tried to restrict it. Protests followed from London, with implied threats of reprisals against American companies, and BP's chairman, Eric Drake, complained to the Attorney-General, John Mitchell. The merger went through. But BP, with its huge hopes for expansion, was now to be a kind of hostage against any British action against American companies.

Exxon, having found its Alaskan oil, was in no hurry to get it out, since they still had ample and far cheaper supplies from Arabia and Iran. The other American sisters were greatly worried that a rush of oil from Alaska would upset their whole balance—particularly Socal, who saw their Californian markets being threatened from the North.[27] Arco and BP were desperate to push ahead fast; but Exxon were moving very slowly. They experimented with sending an ice-breaker the SS *Manhattan*, to force its way through the North-West Passage as a means of bringing the oil through by sea. But BP, though they had to collaborate to 'show they were good Americans' suspected that it was basically a delaying tactic by Exxon to put off building a pipeline. Drake became so impatient that, as he told me, he threatened the chief executive of Exxon, Ken Jamieson, with bringing an anti-trust suit.

Intense opposition to the pipeline was now to come from a completely different quarter. Conservationists on the West Coast discovered that the proposed pipeline would transform the wild life of the whole region; the warm oil would create a wide river in the midst of the ice, thus preventing the migrations of cariboux. The conservationists were strikingly successful, to the growing worry of BP and Arco, who were waiting thirstily for the oil. But Exxon still appeared unconcerned and the suspicion arose in the minds of several BP men: might Exxon be secretly backing the conservationists, as an excuse to delay the production of the oil? The protests of the ecologists set back the transporting of Alaskan oil by four years and the Alaskan alternative was too late to save the West from its dependence on OPEC when the crunch came. It was not until the energy crisis had broken on America that the objections were rapidly overruled.

Over in Europe the North Sea had much more far-reaching implications for the companies and governments. It gradually opened up the possibility of at least lessening Europe's dangerous dependence on the Middle East. But it also began to transform the attitudes of the European governments—particularly of the British and Dutch governments to their two majors, Shell and BP. The interests of companies and governments, which had for so long been assumed to be roughly parallel—to get oil as cheaply and reliably as possible—cut across each other as soon as oil was found in home waters.

The first important prospect in the North Sea had emerged in the 'fifties before the formation of OPEC. Exxon and Shell— the two biggest companies—had been exploring together in Holland, in a joint company. The collaboration was a throwback to the days of the pre-war cartel, when in 1933 Teagle and Deterding had signed a joint agreement for exploration in North-West Europe, and Shell were still bound by the agreement, which they were bitterly to regret: 'we did all the work in the North Sea,' complained a director, 'Exxon got rich on our backs.' In 1959 the joint company found natural gas off the coast of Holland at Groningen—a discovery which was to do more for the economy of Holland than the Dutch East Indies ever did. The first find held the promise of oil, but it was not until 1963 that the North Sea attracted large-scale exploration, for Middle East oil seemed limitless, and discoveries in North Africa seemed safe from nationalism. The European governments involved were very slow to perceive the significance of oil. The Danes even awarded the entire rights for offshore exploration to a single consortium, as if they were Arabs in the 'thirties. The consortium was led by a Danish shipowner. A. P. Møller, and included Shell and Gulf;

having no competition, it was in no great hurry to develop, which eventually led to a major political uproar.

The oil companies were sufficiently aware of the opportunities in the North Sea to establish the legal basis for concessions. They pressed for negotiations which finally led to the Continental Shelf Act of 1964, defining the boundaries of the North Sea. But the British government was very casual and too anxious to get the boundaries settled to argue with Norway. Thus the British were allocated only thirty-five percent of the North Sea, when they might, according to most legal authorities, have obtained a much larger area by taking the issue to the International Court at The Hague.

By 1965 BP had found natural gas off the coast of England. The likelihood of oil now seemed greater, and other majors, and a few independents, were joining in the search. But the British government, which as a consumer had helped the companies drive many hard bargains with the Arabs and Iranians, was now on the receiving end of the companies' pressure. It was Britain as a producer that was in danger of being exploited; and the government was visibly lacking in the expertise, or the will, to supervise the companies.

The first huge areas of the sea, of a hundred square miles each, were leased to the companies as generously as though Britain were a gullible Sheikhdom, with concessions running for forty-six years. And the British government would gain almost nothing in taxes, for the tax system, which Aramco had achieved in 1950 and which the British and other governments had followed, had now given the companies a huge balance of tax credits from their tax payments to the Arabs. The Ministry of Power, who made the arrangements, later explained that they did not wish to be too harsh to the companies lest they encouraged the OPEC countries to be harsh too. But this explanation revealed a curious assumption of Britain's ability to influence the Arabs: as Lord Balogh remarked: 'The Arabs have experts who have forgotten more than the Foreign Office ever knew'.[28] And it also revealed an odd conception of the government's role, that it had a greater duty to protect oil companies abroad than to tax oil companies at home.

More plausibly the Ministry argued that it was essential to attract the maximum possible exploration, as quickly as possible in view of Britain's difficulties with balance of payments. This argument had weight, since the terms could always later be made more stringent. But this was to lead to the accusation against the British government, which they had so often themselves made against the Arabs, that they had broken contracts that were sacred.

It was not until December 1969 that the North Sea at last yielded oil, when Phillips, the American independent, discovered the giant Ekofisk field in the Norwegian sector. The next year BP made the first great British discovery, in the Forties field north of Aberdeen, and a year later Shell and Exxon, still in their giants' partnership, found the Brent field off the Shetlands. The North Sea would clearly now provide a huge prize for the companies, and a whole new economic perspective for Britain. The estimates of the reserves steadily went up, until it became clear that they could make Britain self-sufficient in oil for thirty years, and could supply Norway with far more than her needs.

A hundred years after the birth of the industry, the British, who had so long operated oil by remote control, were now experiencing both its opportunities and its disruptions in the home ground. Texans moved into Aberdeen or London bringing with them the nomadic spirit and the gambling language of oil; referring to Britain as simply an 'oil province', and to the whole North Sea operation as the 'U.K. play'.

Roustabouts for the oil rigs were recruited from the highlands at four times the average wage: Aberdeen became a kind of dour Houston of the North, with the workers from the rigs drinking away their savings while oil company scouts tried to pick up hints of new strikes. The prospect of a local Texas was a huge relief to governments battling to hold up the pound, but the dangers were equally evident. Oil soon began to show the politically divisible effects that it had shown in Canada, Texas or Biafra, fuelling the new Scottish movement for secession or autonomy.

The British were quite unprepared for the industrial opportunities for making the equipment itself. The two British companies, though they had operated abroad for a half-century, were in no position actually to provide rigs, platforms and pipelines. It was not until 1972 that the government commissioned a report on the industrial prospects, when British companies had already lost much of their opportunity. Nearly all the technology and the services were at first supplied from abroad, from Holland, Norway, Japan and specially from Texas. Britain still kept the psychology of being a consumer country, while she was about to become a producer.

The sisters were once again dominant. In allocating the concessions the British government had tried through its discretionary powers to favour both British companies and smaller companies, to avoid the supremacy of the American majors—as the Libyans had done ten years before, and as the Norwegians and Dutch also did. Favoured treatment was given to British groups, including the Gas Council and local syndicates. But the majors with their greater resources for exploration and capital once again emerged as the chief beneficiaries. The British sisters achieved a good share: by 1973 BP controlled twenty percent of the North Sea Oil, and Shell controlled fifteen per cent. But the majority was controlled by the Americans, and the most successful fields included all the familiar sisters, Exxon, Texaco, Mobil, Socal and Gulf, and also the French CFP (Total).

The British government was frightened of antagonising the oil powers, and BP's stake in Alaska had made them more sensitive to reprisals. Even after oil had been discovered they continued to give out concessions on the same generous terms. In the fourth round in 1971 they tried auctioning some of the blocks as an experiment with spectacular results: for fifteen blocks the successful bids totalled £37 million. But the other new blocks were then allocated in the previous way; and the companies still had the apparent prospect of a tax-free bonanza with little benefit to the State or balance of payments. The discovery of oil had caused the same schizophrenia in British governments as had afflicted Washington since the first imports of oil after the Second World War. Were they defending the companies against foreign demands, or defending themselves against the companies? Was it the government or the oil companies which were in control?

It was not until seven years after the first round of concessions that the British parliament eventually caught up with the extent of the companies bonanza. Lord Balogh, who had been economic adviser to the Labour Prime Minister, Harold Wilson, had repeatedly warned that the public was being cheated, and eventually an enquiry was instituted in 1971 by the Chairman of the Public Accounts Committee, Harold Lever. The resulting report had the same kind of historical importance to Britain as the Congressional Report of 1910 on the Standard Oil Trust. Once again, oil was helping to provoke a national counterattack, and the report, published in February 1973, was a powerful indictment of the inadequacy of the civil service in controlling industry.

Its conclusions (allowing for the kid-glove language of such British reports) were devastating. The British Exchequer, reported the committee, would receive a much smaller share of oil revenues than other countries, and British taxes were being preempted by the tax demands of administrations abroad. The Committee was 'surprised' that the Department of Trade had not examined the opportunities for British industry until eight years after the first round of concessions. They advised legislation to increase the tax, and a thorough review of licensing. The new Labour government, returning to office in 1974, duly pushed ahead the changes against angry opposition from the companies — including BP which was now vigorously campaigning against the government which owned 48 percent of its shares.

The government of Norway, in the meantime, after the discovery of the great Ekofisk field, took a much sterner line with the companies. They first followed the British in their discretionary system of handing out concessions, but became much tougher with the companies as soon as oil was discovered. The reason was clear: having no international oil companies, the Norwegians had no reason to sympathise with them. At a time when OPEC was becoming more vocal, they were tempted to see their own country, with its small population and sense of separateness from Europe as having much in common with OPEC.

As the OPEC countries became more interested in conservation of their resources, so the Norwegians were determined not to extract their oil too fast, and to resist pressures to export rapidly to the rest of Europe. They encouraged their own industrialists to supply equipment and formed their own company to make oil rigs. They kept in touch with the OPEC experts, and particularly with Yamani. The Texans referred to Norwegians sourly as the 'blue-eyed Arabs'.

The hopes that the North Sea, and Alaska, would rescue the companies from their dependence on OPEC had been partially fulfilled, but with an ironic twist. For the companies, coming up against their home governments, were to find that they, too, had learnt something from the Arabs.

NOTES

1. Benjamin Shwadran: *The Middle East, Oil and the Great Powers*, (revised edition) New York, 1973, p. 536.

2. Wanda Jablonski in *Petroleum Week*, July 22, 1960.

3. Edith Penrose: *The International Petroleum Industry*, p. 195.

4. See Chapter 6.

5. See Chapter 5, p. 124.

6. Zuhayr Mikdashi: *The Community of Oil Exporting Countries*, London, 1972, p. 33.

7. David Hirst: *Oil and Public Opinion in the Middle East*, London, 1966, p. 112.

8. Benjamin Shwadran: *The Middle East, Oil and the Great Powers*, New York, 1973, p. 516.

9. James Akins: 'The Oil Crisis', *Foreign Affairs*, April 1973.

10. Shwadran: pp. 503–538.

11. Interview with author: February 1975.

12. Richard Rovere: *The American Establishment*, New York, 1962, p. 11.

13. Multinational Hearings: Part 6, p. 290.

14. Multinational Hearings: Part 5, pp. 255–7.

15. Multinational Hearings: Part 7, p. 309.

16. Interview with author, September 1974.

17. See Chapter 4, p. 84.

18. See Multinational Report: p. 101 and Chapter 6, p. 137.

19. Multinational Hearings: Part 7, p. 309.

20. Multinational Hearings: Part 7, p. 304.

21. Multinational Hearings: Part 8.

22. Multinational Hearings: Part 7, p. 285. See also Chapter 6.

23. Multinational Hearings: Part 7, p. 257.

24. Interview with author, September 1974.

25. Multinational Hearings: Part 7, p. 720.

26. Multinational Hearings: Part 9.

27. Socal Memo from S. E. Watterson, December 6, 1968. Multinational Hearings: Part 7, p. 360–3.

28. The North Sea Blunder: *The Banker*, London, March 1974.

9

Sisters Under Stress

If this international institution did not exist it would be necessary to invent one.

J. E. Hartshorn, 1962

By the end of the 'sixties, in spite of the opposition of OPEC and the competition from intruders, the seven sisters were still the dominant powers in world oil. Between 1960 and 1966 their share of oil production outside North America and the Communist countries, had actually gone up from 72 to 76 percent, leaving only 24 percent for all other companies.[1] Their profits, whatever their protestations to OPEC, and despite the falling prices, were still huge compared to most other industries. The rate of return of most of them was higher in 1966 than it had been in 1960,[2] and the companies were able to finance most of their exploration, their tanker fleets, refineries, tank farms, trucks and filling-stations, together with their expansion into the lucrative new business of petrochemicals, out of the profits of their crude oil abroad. The companies could and did argue with OPEC that this expansion was essential to create a market for the oil which would otherwise be unsellable. But the remarkable fact remained that these seven had built themselves up into some of the biggest corporations in history primarily through the ownership of concessions in developing countries, and predominantly in the Middle East.

They gave every appearance of permanence and stability, with their self-perpetuating boards and bureaucracies and their evident ability to survive two world wars and countless revolutions across the continents. Their engineering achievements commanded the awe of governments and publics. Great refinery complexes rose up along the coastlines, with their grotesque skylines of strange shapes—spheres, towers, and cylinders interwoven with twisting pipes and surrounded by white tanks—which looked like giants' kitchens which had outgrown any human fallibilities. They seemed to mark the triumph of technology over man. The whole style of the corporations, grown smoother and more confident over the decades, suggested a lofty superiority to all governments. It was hard to remember, behind all this grandeur and hardware, that they all rested, like upside-down pyramids, on the most perilous political base.

The anatomy of these strange industrial organisms has often baffled the economists, as well as the politicians, who looked into them. J. E. Hartshorn, analysing them in 1962, had described how they appeared to live on an imaginary island—'the Shell Company of Atlantis'. The companies have frequently seemed to inhabit a no-man's land between the defined areas of governments and business—an impression increased by the strange life-style of the oil executives, who seem hardly to touch ground between their international adventures. The detachment from national governments has, as we have seen, been an important part of the political justification of the companies, as the 'buffers' between nations, performing what Hartshorn calls 'the business in between'.[3] They can make their long-term investments across the oceans with both greater resources, and greater insulation from political pressures, than the national governments at both ends.

The Western governments, in Washington, London or The Hague, largely accepted the advantages of this detachment. In Washington, as we have seen, it suited the State Department to separate the foreign policy of the oil companies from their own, particularly concerning Israel. In the Foreign Office in London, the diplomats at the 'oil desk' were told not to interfere with the commercial policies of the great oil companies, which did such wonders for the balance of payments; and the disaster of the Suez adventure was an object-lesson in the dangers of governments trying to intervene without being asked. Shell were able to point out that they were able to do business in Cairo soon after the Suez War, while the British diplomats were still thoroughly *non grata*. Likewise after the Six-Day War in 1967, American oil executives could boast that they were still very well received in the Arab countries, while diplomats from the State Department were quite unacceptable. Even in Paris, where the French government had long regarded the chief oil company, CFP, as its chosen instrument, the oil executives became increasingly separate from the Quai d'Orsay.

The appearance of neutrality encouraged the oil companies to treat governments cavalierly, and to present themselves as the guardians not only of the security of the West, but of the peace of the world. Yet this neutrality was always illusory: as soon as they were in fundamental difficulties, like Aramco in wartime or BP in 1951, they would fly to their governments for help. And despite the companies' posture as the fearless champions of adventurous free-enterprise, it was sometimes the governments who pushed them into their exploits: as the State Department pushed the companies into Iraq in 1920, or into Iran in 1954.

Looking back on the history of their industry, American oilmen are prone to use the well-worn simile, 'It was like Topsy, it just growed'. Certainly the growth of the industry was so rapid, so unpredictable and accidental that it was beyond the control of any man or group of men to plan it or circumscribe it. The black fluid seemed often to be itself the real master, spurting up and subsiding from bleak corners of the world, teasing the West by its combination of indispensability and maddening inaccessibility. But the companies were not quite like Topsy. From the beginnings, when the chaos of Pennsylvania led to the order of Rockefeller, the industry was controlled, first by one master, then by several, taking very deliberate decisions. And later the Western governments, though reluctant and often incompetent, intervened at a few decisive moments of history to change the whole balance of the business—as they were soon to intervene once again.

Between times, inhabiting their no-man's land, the companies could claim with some justification to have become uniquely internationalised, with their world-wide affiliates, their polyglot managers, and their consciousness of politics everywhere; Exxon could boast of being the first multinational, and Shell of its cross-postings between nations. But the fact remained that their shareholders were predominantly American and British; and their boards represented only their own country—and a small segment of that. They may well have believed that they were dispassionately representing not just their own interests, but the world's; but the whole environment in which they worked and lived was dominated by the need for home profits, with minimal taxation. The rapidly changing face of the developing world had found very little reflection in the upper hierarchy of the great companies which had elected itself over the decades. As Professor Penrose put it in 1968: 'The deeper root of the problem is simply that international firms, including the oil companies, have not yet found a way of operating in the modern world which would make them generally acceptable as truly international institutions.'[4]

And however much the more enlightened company-men might see themselves as carrying the burdens of the whole world, they were not accountable to any body that could judge their performance. As Rockefeller and Standard Oil in the nineteenth century had become bigger than individual states of the Union, so these giant companies had become larger and richer than most national governments. They had romped across the world, ahead of most other multinational corporations, and way ahead of any effective international authority or regulation. For some observers, this simply meant that they were the forerunners of world government. They would stimulate global organisations as a counterweight to their own power, rather perhaps as Rockefeller had stimulated the Federal counterweights in Washington a century before. But such an organisation was clearly a long way in the future. In the meantime the great companies—if they were serious about their unique international burden—had a unique responsibility to provide their own counterweights, to show themselves honestly to the rest of the world, and to play the part of economic statesmen, as well as pursuers of profits.

EXXON

The seven companies shared these same international opportunities and limitations, as they had grown up together, with their network of joint ventures and consortia closely interlocked. But each of them saw the world with a different perspective, influenced by its own needs to sell or buy oil; and each still bore the marks of its own history, and lived partly in 'the long shadow of its founder'.

Two of the sisters, Exxon and Shell, still dominated the world's oil as they had done for the past fifty years. In the oilfields they kept on coming together in their global quadrille, whether in the Iranian Consortium, or in the North Sea. But at the selling end they competed with visible intensity, the yellow signs of Shell and the red signs of Exxon or Esso staring at each other from filling-stations across the world. They still in some respects had opposite outlooks, dating back to their very different sources of oil.

The organisation that Rockefeller established a century before had long ago expanded into a self-perpetuating managerial corporation. The three thousand senior executives were systematically recruited from the universities, earmarked and watched for promotion: they climbed up through the escarpments of managers and 'co-ordinators', towards the plateau of the Main Board, and the five-man Executive Committee. Planning was no longer a question of hunches. Forecasts for long-term supply and demand were compiled each year in the Exxon Green Books. Once a year, in October, came the exhaustive budget review, which involved more money and longer forecasting than the budgets of most countries where Exxon operated. First the Economics and Planning

WORLD'S 12 LARGEST MANUFACTURING CORPORATIONS RANKED BY ASSETS IN 1972

Rank	Company	Assets ($000)	Sales ($000)	Rank
1[1]	Exxon	21,558,257	20,309,753	2
2[1]	Royal Dutch/Shell	20,066,802	14,060,307	4
3	General Motors	18,273,382	30,435,231	1
4[1]	Texaco	12,032,174	8,692,991	10
5	Ford	11,634,000	20,194,400	3
6	IBM	10,792,402	9,532,593	7
7[1]	Gulf	9,324,000	6,243,000	12
8[1]	Mobil	9,216,713	9,166,332	8
9	Nippon Steel	8,622,916	5,364,332	17
10	ITT	8,617,897	8,556,826	11
11[1]	BP	8,161,413	5,711,555	15
12[1]	Socal	8,084,193	5,829,487	14

1. Indicates one of the seven sisters. Source: *Fortune* Magazine, May and September, 1973.

Department gave its overview, country by country. Then the functional co-ordinators gave their budget breakdowns; followed by the production co-ordinator, the refinery co-ordinator, the transportation co-ordinator and the marketing co-ordinator. Finally the Investment Advisory Committee weighed up the demands, and passed its findings to the Executive Committee, who decided.

The chemical engineers still came up through the refineries to join the board or become chief executives, in a steady progression. Monroe Rathbone gave way in 1965 to another engineer, Michael Haider, and Haider gave way in 1970 to Ken Jamieson, the Canadian from Medicine Hat, who was now really an honorary Texan.

The rule of the engineers was not surprising in an industry which invested so hugely in hardware—rigs, platforms, refineries, pipelines. But their overwhelming predominance on the Exxon boards—as on other oil boards—was also a reflection of the distorting effects of taxation. Ever since the oil companies had achieved their relief from U.S. taxes in 1951, on top of the earlier relief through the depletion allowance, they had adjusted their internal accounting to take as much profit as possible 'upstream'—from producing crude oil abroad—on which they paid no U.S. taxes; rather than 'downstream', from selling oil products to customers. And they were easily fooled by their own accounting into supposing that downstream did not matter.

Selling oil was regarded less as a source of profits, than as a problem of finding 'outlets': the attitude was revealed in the word itself, and in the multiplication of filling-stations, duplicating themselves along the freeway ('they seem to be born in litters', complained Harold Ickes). The indifference to 'down-stream' profits was reflected on the board, where the marketing men lost caste over the engineers, with much less say on policy decisions (as in the disatrous 1960 decision to reduce the posted price). The ascendancy of the engineers seemed assured so long as the profits from crude oil were secure. But once they were threatened by OPEC, the imbalance became very dangerous; for the companies desperately needed to find profits downstream, with little expertise to help them. It was one of the oddest facts about these giants, with all their resources and commercial instincts, that they were not very good at selling oil.

But the more serious and evident weakness of the Exxon engineers was their lack of international perspective. Only one member of the board, 'Pete' Collado—who had been Acheson's economic adviser at Bretton Woods—had experience of international diplomacy; and after the retirement of Howard Page in 1970 Exxon had no real oil statesman. The self-perpetuation of the board, their separation from their shareholders—including the Rockefellers—made the possibility of adapting more difficult. Their character—as Richard Funkhouser had warned twenty years before—was becoming dangerously like a dinosaur's. As Exxon became more exposed to politics, both in the Middle East and Washington, so its lack of world outlook became more evident. As one former Exxon executive, now in government, put it to me: 'They won't solve their problems until the Rockefellers or the

Chase Manhattan intervene, to put in a chief executive who understands world politics.'

Exxon, like many other multinationals, is full of the rhetoric of global responsibilities; it likes to stress that it serves not only its American shareholders but all the nations where it operates. In some respects an international outlook is forced upon them, for the heads of their big subsidiaries abroad, of Esso Europe or Esso Japan, have considerable bargaining power. But their global outlook is severely limited, not only by their shareholders, but by the narrowness of their board, and their ultimate sense of dependence on the U.S. government. Exxon have seen themselves as spreading American ideals, as they revealed in a famous passage in their annual report in 1962: 'The public statements made by our managements, our written communications, and our advertising seek to emphasise the benefits of free competitive enterprise and private international investment.'

Exxon were still inclined, when in trouble abroad, to look for help to Washington, to wield the 'big stick'. Their clumsiest blunder was in their dealings, not in the Middle East, but in Peru. Their subsidiary there, called the International Petroleum Company, or IPC, had been in dispute with the Peruvian government ever since 1918, when they cut off oil supplies through their tankers to enforce their terms. By the 'sixties, with the new government of Belaúnde, the argument became more bitter, with the Peruvians insisting on a share in the company and back payments of taxes. Exxon stood firm, and invoked the support of the State Department, who in 1964 began withdrawing aid from Peru for two years to exert pressure. Washington later relented, but in the meantime the IPC issue had become more explosive, taken up by the left-wing opposition. When eventually in 1968 Belaúnde announced a settlement over the IPC question, the issue helped to trigger off a military revolt, and the coming to power of the new Junta, which swiftly nationalised the IPC properties. Exxon's alliance with Washington had done nothing to help their own interests; and had served to discredit the Peruvian elite.[5]

In the Middle East, Exxon had come to be regarded as the leader of the American companies or, when the consortia were in dispute, as the mediator between them: Ken Jamieson saw himself in a more conciliatory role than his predecessor, Michael Haider. Exxon had now been involved in the Middle East for fifty years, had seen Kings and Sheikhs come and go, new countries

invented, economies transformed. With their continuous history,
and their great resources, they were in a unique position to take the
long view of the future, to prefer long-term stability to short-term
quick profits. But in the events in the following chapters there was
little sign of such statesmanship. When they saw the age of concessions coming to an end, they could not resist hanging on to them
till the last moment, with their huge profits. When threatened,
they ran once again to Washington.

MOBIL

The old Rockefeller companies are still sometimes called by
their critics 'The Standard Oil Group'; and the Rockefeller
family still owns shares in the three Standard sisters: 2 percent
in Exxon, 1·75 percent in Mobil, 2 percent in Socal. But they have
each developed in different directions.

Mobil was for long regarded as Exxon's little sister, dependent
on her bigger rival both for advice and for oil—for Mobil had
always been hungry for crude. In the pre-war years their association was close: for thirty years from 1930 the two companies had a
joint subsidiary called Stanvac which sold oil in fifty countries
abroad, until it was broken up by anti-trust action in 1960. But
since then Mobil has become apparently independent, and in 1966
it dropped the last hint of its Standard Oil origins, changing its
name from Socony-Mobil simply to Mobil. And it has emerged as
the most aggressive, and in many ways the most sophisticated, of
the American sisters; proudly associated with New York, much
concerned with communications and image, subsidising the TV
'Masterpiece Theatre' and advertising relentlessly about the problems of oil.

In 1969 Mobil acquired a chairman and chief executive,
Rawleigh Warner, who was determined to assert Mobil's independence. He is an aristocrat of oil, with the looks of an old-
fashioned film star and a Princeton education. His father was the
head of an independent oil company, Pure Oil, and he first joined
another independent, Continental, when it was expanding abroad,
before he moved over to Mobil. He astonished the oil world by
appointing as President a man from right outside the old Wasp
tradition of the oil boards; a lawyer-accountant called Bill Tavoulareas, the son of a Greek-Italian butcher from Brooklyn, whose
name few oilmen could pronounce. 'Tav' was already a phenomenon within Mobil; an irreverent, fast-talking numbers-man

who had the crucial Rockefeller talent for lightning mental arithmetic. Like other Greeks in the oil business he had an instinctive
global awareness.[6] He was a forthright Brooklyn boy, outspoken
and impatient with the slow style of the company. As Mobil became more dependent on the Middle East consortia, particularly
Aramco, so Warner decided that there was only one man in Mobil
who could really understand the accounting well enough to hold
his own with the Exxon expert, Howard Page: he thus promoted
Tavoulareas to be Middle East negotiator. Tav would then explain
to his Mobil colleagues, when battling with Exxon: 'This is what
they want, this is how they'll try to get it, and this is how we'll stop
them'; and he did.

Tavoulareas is now widely regarded as the ablest of the major
oilmen. He sits in shirt-sleeves, talking at top speed, blinking,
twitching and staring, running to the telephone like an imp let
loose; saying gimme and lemme, whadda ya want. In the Middle
East, Tav was determined to increase Mobil's share, and was
much more prepared to consider new partnership arrangements,
which antagonised the other sisters, but also brought him closer
to the producers; and he formed a close friendship with Yamani
in Saudi Arabia. But Mobil was always in the minority, and while
it was more open-minded than most, more inclined to explain its
problems, it lacked any really far-sighted policy about the future
of oil.

GULF

Over in Pittsburgh, the headquarters of Gulf convey the unique
character of the company, a huge self-contained family firm. The
great stone skyscraper with its high lobby in the grand style of the
late 'twenties, like a tomb, evokes a sense of calm permanence.
Beside the entrance is an office of the Mellon Bank, and all round
it other skyscrapers have risen up—the aluminium tower of Alcoa,
the black criss-cross shape of U.S. Steel, the palace of Koppers—
all Mellon companies. The word Mellon is blazoned round the city
from banks, from a gallery, from a university—while on the freeways the orange Gulf signs outnumber all other names. The scene
makes the point; that Pittsburgh, Gulf and the wealth of the Mellons have grown up together, since the family first financed Gulf
seventy years ago.

The Mellons kept much bigger holdings than the Rockefellerss
and they still rival the Rockefellers in wealth. By 1973 the Mellon,

still held over 20 percent of the shares in Gulf—an astounding chunk of the tenth biggest American corporation—and there were two representatives on the board: Jim Walton, a grandson of William Larimer Mellon, and Nathan Pearson, the family's investment adviser. The Mellon influence may have been critical in achieving Gulf's biggest coup, when Andrew Mellon helped to get half the oil from Kuwait. Andrew's nephew, Richard King Mellon, 'the General', later dominated Pittsburgh and kept a close eye on the companies (when asked what he did for a living he said, 'I hire Company Presidents'). But his sons are not interested, and Paul Mellon, now the head of the family, long ago fled to Virginia, where he hunts and collects English water colours.

Gulf still has many Pittsburgh qualities, including its proud conservatism. But the real roots of the company management are not in Pennsylvania but in Texas: in Gulf too the top executives have nearly all come from the South West. E. D. Brockett, who became chairman in 1965, began as a roughneck in Crane, Texas; Bob Dorsey, who succeeded him, was a chemical engineer at the University of Texas; Jimmy Lee who became President under him, began his career at Port Arthur. And the sources of the company's great wealth have been first Texas and second Kuwait. Since Gulf first struck oil in Kuwait in 1937 they floated to greater and greater profits on its oil, hectically trying to find new outlets for the flood. The Pittsburgh executives could never quite come to terms with this strange alliance with a tiny territory at the other end of the world. They took most of their political advice from their partners BP, who gave them patronising lectures on how to deal with the Arabs. They tried to escape from their perilous dependence. They invested heavily in Angola only to find it a much more dangerous territory, a battleground between black and white. They moved into coal and nuclear energy; they bought an insurance company (CNA), an industrial centre in Florida, and a whole new town outside Washington called Reston. But they never had another bonanza to compare with Kuwait, and in the meantime, as we will see, they were caught up in political scandals over their bribes to foreign governments.

SOCAL

Two of the American sisters, Socal and Texaco, have remained bracketed together in the minds of the others for the past forty years, as the outsiders of the industry, the terrible twins, the

people who always say no. Their close association began when
Socal invited Texaco to join them in their Saudi-Arabian adventure in 1936, and they formed a joint marketing company called
Caltex, to transport and sell the oil through the world—until in
1967 the joint company was broken up in most of Europe. The
two companies, with their long experience together and with their
Far-Western attitudes still have much in common, most of all a
stubborn resistance to change: though like many hawks, as we will
see, they have tended to switch suddenly from total intransigence
to total capitulation.

Over in San Francisco, Socal makes a positive cult of conservatism: 'This', one of the officials boasted to me, 'is a *very* stuffy
company.' On the eighteenth floor, the directors are served by
reverent black flunkeys and timid secretaries: it is only in the last
few years that Socal has allowed women secretaries. The board has
been peopled with engineers from California and Texas, and much
of the company's business has always been done by the local law
firm of Pillsbury, Madison and Sutro, whose former senior partner,
James O'Brien, an anti-trust expert, now sits on the board. An
important influence on the company, too, has been the former head
of the CIA, John McCone, a shareholder in Socal, whose relationship with government has been shrouded in mystery. The chairman from 1966 was Otto Miller, a chemical engineer from
Michigan who had planned the great Ras Tanura refinery in Saudi
Arabia, and later took charge of the Eastern Hemisphere. A stolid
Republican and backer of Nixon, he was always suspicious of the
liberals in the East.

Behind all the confident Californian façade, the company's
prosperity has been built on a single country 10,000 miles away.
Ever since it first found oil in Saudi Arabia, its dependence has
steadily increased, so that by the 'seventies half its oil came from
its share in Aramco. Once it flowed, the main problem was to find
outlets, and to minimise taxes by intricate arrangements of transfer
pricing within its subsidiaries, at which Socal is specially expert. In
all its dealings abroad, it had been obsessed by this Arabian Jack-
pot, reluctant to share it with any new partner, whether from the
West or the East. In the late 'sixties George Ball, the former
Democratic Under-Secretary of State who became a director of
Socal, advocated that in view of the political dangers of the all-
American ownership of Aramco, European companies should be

permitted a share. But Miller would not consider it, and soon afterwards Ball—who was also very critical of President Nixon—was not reappointed a director.

TEXACO

'We all hate Texaco'; said an Exxon man. 'If I were dying in a Texaco filling-station,' said a Shell man, 'I'd ask to be dragged across the road.' The other sisters are united in their resentment of Texaco, and it is not hard to see why. Texaco has always taken pride in being the meanest of the big companies, the loner in the Western, refusing to contribute anything except for profit (apart from their patronage of opera), while their return on capital has been consistently the highest. The flag of the skull-and-crossbones, which their founder, Joe Cullinan, flew over his offices still seems an apt emblem.

They are not in fact more Texan than the others. Ever since Joe Cullinan was ousted in 1913, they have been run from New York, where they now inhabit part of the Chrysler skyscraper. From there they established a tradition of skinflint management and centralised control which was reinforced by the man who dominated it for two decades: Gus Long, a granite-faced salesman from Florida, joined the company through his first father-in-law, who was Cullinan's attorney, and he soon made his name as a 'no'-man. He resigned as chief executive in 1964, but was brought back six years later: he is still a director on the executive committee, at the age of seventy-five, and the board is full of Long's men.

Former Texaco men lovingly exchange stories about the company's penny-pinching, totting up tiny expenses, or refusing to allow their experts to contribute to industry meetings. In Libya, it was said, when Texaco cabled the revolutionary government announcing a new price involving millions of dollars, they cabled by the cheap night-letter rate. Texaco, with its selfishness and greed, is a convenient candidate for the role of the baddy, the more so with its growing international involvements: in selling its Arabian oil, it has become much more politically visible in Europe; in 1956 it bought out the Regent Oil Company in Britian, and in 1966 it bought one of the biggest German oil companies, Deutsche Erdöl. Some oilmen maintain that without the negativity and lack of foresight of Texaco and Socal, the whole industry could have made more imaginative moves in the Middle East.

But it is also argued that Texaco is the most honest and the least self-deceiving of the companies, that its parsimony simply shows responsibility to its shareholders. It has never pretended to be anything more than it is: a concern for making money, as quickly as possible, not a benevolent institution for world peace. If the politicians want that, is that not the government's responsibility? The case of Texaco raises in its extreme form, the question which runs through this book; what do we really expect oil companies to *be*?

SHELL

: Across the Atlantic, Royal Dutch/Shell has presented itself in a much nobler light than Texaco, or any of the American sisters. More than any company it has seen itself as an international institution, and neither in Holland nor in Britain has it been fiercely attacked by anti-trust or other critics, as have Exxon and the rest. Since the wild days of Deterding it has taken precautions that it should never again be controlled by a dictator, and has erased 'that old bandit' (as one of the current directors called him) from its corporate memory. The Group is run by a committee of seven or eight, and each chairman is ruthlessly retired at sixty.

Since Deterding's departure, after a period of demoralisation, a succession of respectable chairmen inspired the confidence of their home governments. John Loudon who was chairman for nine years until 1965, was a Dutch Jonkheer and a British knight, with the style of a Prince of the Industry, patiently suffering the insults of the populace: 'I don't know why,' he said to me in 1962, 'but oil always seems to have a smell to it . . . the dogs may bark, but the caravan moves on.' His successor for two years was a Dutch engineer, Jan Brouwer, and he was succeeded by a British aristocrat, Sir David Barran; the son of a former head of Shell's Venezuelan company, he added to the impression that the Group was a public service. Both in the Hague and in London, the great company was regarded as an indispensable part of the nation's economy, a steady contributor to the balance of payments, and the Treasury always made special exceptions for Shell in enforcing exchange control. The public image was duly fostered by patronage, elegant Shell guidebooks, and the slogan 'You can be sure of Shell'; a stockbroker's rule was 'Never Sell Shell'. The British Shell executives, mostly graduates from Oxford and Cambridge, were inclined to talk about oil not as a profitable fluid, but as part of a public duty; as one rebel complained to me: 'I wish they'd stop talking about it as if it was a bloody faith. Anyone would think it was a church.'

And Shell were able to present themselves, with more justification than the other companies, as being genuinely international. The Anglo-Dutch condominium, which had been in rivalry under Deterding, had settled into peaceful co-existence, making both sides more tolerant of other foreigners. Shell's long dependence on foreign oil forced them to be more attentive than Exxon to susceptibilities abroad. The shock of the Mexican nationalisation had made them more subtle in their dealings with Venezuela, their most crucial source of post-war oil. It had induced a careful programme of 'regionalisation', trying to buy in the local elites throughout their world empire—presenting Deutsche Shell, Shell Italiana, or Shell Senegal as essentially local companies. Despite its ponderous headquarters, Shell men insisted that 'the Group' was really a federation of modest companies, which just happened to have a committee of managing directors who were English or Dutch. In Shell Plaza in Houston, the local directors of the U.S. subsidiary (which is 30 percent American-owned) like to emphasise that they are wholly autonomous.

Shell certainly appears more open-minded than the others, more prepared to discuss political problems, less confident of its ultimate rightness. Its need to buy oil from others, ever since its beginnings, has forced it to live more continually on its wits, and to balance its client countries. Unlike Exxon, it has never been able to retreat into self-sufficiency, back to a domestic stronghold. Yet behind the picture of the global federation there has always been a very large qualification. It is not only that Shell remains very firmly Dutch-English, preoccupied in the end by its home shareholders and governments, still treating everyone, as Churchill complained, with 'the full rigour of the market'. It is also that Shell, having had a greater opportunity than the others to adapt itself more genuinely into a new kind of international corporation, to involve the producing countries and their leaders more deeply, has not really faced up to it. In the events in the following chapters, Shell can claim to have been the most far-sighted of the seven, and to have shown more responsibility than the others. But that could be expected from its uniquely international genesis. With its long global experience, it was still painfully slow in catching up with the changing face of the world.

BP

In the City of London a new skyscraper was completed in 1967 which surprised even other oilmen with its extravagance. With a high lobby with marble pillars, acres of panelled conference halls, a self-contained village of canteens, shops and recreations, and a garage beneath with a fleet of company Jaguars. Above the entrance was the proud name, Britannic House, and in front was a wide piazza with a flagpost flying the green shield of British Petroleum. In the middle of the piazza a metal object like a modish sculpture revealed the source of much of this conspicuous wealth; an old 'Christmas Tree' well head from the Iranian oilfields, which had first flowed with oil in 1911, and which had carried a total of fifty million barrels.

In the sixty years since, it was the oil from Iran and the Persian Gulf which had built BP into one of the world's biggest companies, and financed its successes in the North Sea and Alaska. The British government's half share was little evident in its commercial dealings, yet there was no doubt that BP had a very special sense of patriotic confidence among the sisters. The company still raised the awkward question: who is an oil company responsible to?

Inside the building, the answer is not immediately clear. On the one hand BP, in its relationships abroad and with other companies, is determined to appear just another international company: it carefully uses only its initials to disguise its Britishness, BP men indignantly deny that they lean on the support of the British government, and like to mock the incompetence of politicians. At the same time their whole regimental pride, and also their low profits—consistently the lowest return on capital of the seven— appear to be linked to the sense of government service. The board has the look of a miniature House of Lords. In 1974 the fourteen directors included six peers, including (as one of the government directors) the former head of the Foreign Office, Lord Greenhill; an ex-head of the Bank of England, Lord Cobbold; an ex-chief of the Defence Staff, Lord Elworthy; an ex-ambassador to Moscow, Lord Trevelyan. The BP senior staff has always been peppered with ex-military men and ex-government officials. It is the kind of meeting of government and industry that would leave the trust-busters from Washington gasping.

At the head of this dignified board, a new chairman was appointed in 1969 who expressed well enough the confident spirit.

Eric Drake (soon Sir Eric) had come in as an accountant and a Cambridge rowing-blue: he was general manager in Iran at the time of the Mossadeq crisis of 1951. He appears to be afflicted with few doubts about his role: he puffs a pipe, puts his thumb in his braces, and talks with tolerant amusement about the problems of other companies. For his part he can always get in to see the Prime Minister when he wants; Britain is the only country with sensible relations between government and oil companies. The British government sometimes helps with possible concessions, and BP in turn keeps them informed. But the government agrees not to interfere in commercial decisions, which actually make BP less vulnerable to government than many private companies. Sir Eric regards the British government with affectionate contempt for their ignorance of oil: 'to give them their due', he told reporters in 1975, 'they did *try* to understand the problem'. Sir Eric's attitude, and that of other BP officials, seems all very persuasive. Why don't Americans realise that all this conflict between government and companies is wasteful, that they all have the same interest?

To outsiders it might look as if these British oilmen and diplomats—both with their own Oxbridge education, their Kensington addresses—share the same perspective; but the reality is otherwise. The government officials are still inclined to regard BP, as they did in 1951, as a kind of Frankenstein monster. They complain that BP's pride is really based on an elaborate self-deception; that the company cannot face up to its dependence on its colonial origins and defence agreements. It pretends to the government to have an independence it doesn't have, and having got into trouble comes running home, too late. The diplomats are elaborately briefed by the company, but BP does not take advice from the Foreign Office. And in the meantime the Treasury men have periodically tried, without success in spite of their being half-owners,[7] to extract the truth about BP's sources of profit. The relationships with both departments were soon to be much more strained in the oncoming crisis. On one front, in the North Sea, which brought the question of taxation to a head; and on the other, in the critical confrontations with OPEC which pressed home the point that oil was too important to be left to oil men.

THE LOBBY

By the early 'seventies the oil lobby in America was a less formidable and solid political force than in its heyday in the 'fifties. Sam Rayburn, the Texan Speaker of the House, who had been one of oil's greatest champions, was long dead, and Lyndon Johnson was retired. Even Wilbur Mills was before long to become discredited by his relationship with the 'Argentine firecracker'. The new generation of Congressmen were more representative of the big-city voters from non-oil states, so that the chief lobbyist of Shell Oil, Carter Perkins, could complain to the American Petroleum Institute: 'no longer can we rely on Congressmen in leadership positions to recognise and take into account the seriousness of proposed legislation on the petroleum industry . . . it may have been foolish for our industry ever to have relied on a few knowledgeable and influential Congressional leaders.'[8]

There was now little hard evidence of an effective conspiracy of oil interests, as there had been forty years before at Achnacarry Castle. The chief meeting-place of the industry, the American Petroleum Institute (API), which had been described in 1934 by Clarence Darrow as 'the switchboard for the contolling companies', was now muddled with crossed lines; it recorded all the discords between the majors and the independents, disunited in their attitudes to foreign oil. It was true that the relationship between the majors in their production abroad was sufficiently close to regulate with great subtlety the supplies of oil in ten nations: they moved in parallel, watching each other anxiously in their common preoccupation to avoid an uncontrolled glut. But the cosy arrangements were already being disturbed by the rude invasions of the independents, and it was hard to define them legally as amounting to a conspiracy. In the words of one recent radical study of oil interests:

These relationships are not perfected in a group of cigar-smoking individuals conniving in unison in a smoke-filled back-room. If any arrangements exist, this study has no knowledge of them. Rather, it is joint agreement reached by gentlemen who think alike, business leaders whose protection of and concern for one another makes an instinct for self-survival, people and institutions who through very reciprocal favors, men of substance interested in the preservation of wealth and power, and deft operators whose very least operational techniques would be to pursue an objective frontally with full disclosure.[9]

THE SEVEN SISTERS' SHARES OF WORLD CRUDE OIL PRODUCTION—1972

Company	Production in U.S. (Thou. b/d) (1)	% of total U.S. production (2)	Production in Middle East[2] & Libya (Thou. b/d) (3)	% of total M.E.[2] & Libya production (4)	Production in all OPEC (Thou. b/d) (5)	% of total OPEC production (6)	Production[1] world-wide, (excluding E. Europe & China) (Thou. b/d) (7)	% of world production (excluding E. Europe & China) (8)
Exxon	1,114	9·9	2,527	12·9	4,050	15·2	6,145	14·7
Texaco	916	8·1	2,155	11·0	2,674	10·0	4,021	9·6
Socal	528	4·7	2,155	11·0	2,614	9·8	3,323	7·9
Gulf	651	5·8	1,887	9·7	2,409	9·0	3,404	8·1
Mobil	457	4·1	1,178	6·0	1,477	5·5	2,399	5·7
BP	–	–	3,903	20·0	4,506	16·9	4,659	11·1
Shell	726	6·5	1,372	7·0	2,877	10·8	5,416	12·9
Total	4,392	39·1	14,165	77·6	20,607	77·1	29,367	70·0

1. Taken from company annual reports.
2. Excludes Bahrain.
Source: Multinational Hearings: 1974, Part 4, p. 68.

While the companies had no obvious get-togethers, and while many of the independents—like Getty and Hunt—had never actually met each other, they shared the incentives and interests of the self-contained world of 'oildom'—a kingdom which had extended far abroad since the days of Rockefeller, but which seemed paradoxically to be more self-contained as it was more scattered. For the more oilmen travelled, the more they were caught in each other's company in desert outposts where oil was the chief common subject. The consortia of companies in the Middle East provided a legitimate meeting ground between the companies, particularly the Iranian Consortium which included all seven and several independents, which had its meetings in London. London—a traditional haven from anti-trust—was in many respects the capital of the international industry: oilmen could move easily from the Riveroaks Club in Houston to the River Club in New York to Les Ambassadeurs in London. There were many Anglo-American intermarriages, like the Pages or the Strathalmonds, and the world of oil was still—with only a few exceptions like Tavoulareas of Mobil—emphatically Wasp. There were still very few Jews in the business, even in Shell which was founded by a Jew; and the old Wasp tradition had been reinforced by new pressures from the Arab states—particularly on Aramco, whose New York office excluded Jewish employees.

The common interests of the oilmen had been bound together by a network of communications, banks and accountants specialising in the arcane science of oil accounting and taxation. The Morgan Guaranty Bank provided underwriting for most of the sisters—Exxon, Mobil, Shell Oil and Texaco—while the Chase Manhatten and the First National City provided much of their international services. The chairman of the Chase, David Rockefeller, has been the subject of great curiosity in the oil business, as the grandson most actively involved in finance. He does not sit on the board of any Standard Oil company, but he and his family still have holdings in all the Standard sisters; and Rockefeller himself has close contacts with the Middle East, which he visits frequently. The figure of John McCloy stands at the crossroads of several different Rockefeller worlds—the oil companies, the Chase Manhattan and the personal interests of the Rockefeller brothers represented by his law firm. All oil companies have interlocking directorships with banks and investment houses, and the great oil foundations bequeathed by the founders of the industry have linked the prosperity of the companies to the world of philanthropy. In 1971 the Rockefeller Foundation and the Rockefeller Brothers Fund together owned $250 million in Mobil; three Mellon foundations between them owned shares worth $435 million in Gulf.[10]

Since tax-avoidance is the critical element in the oil companies' financial power, the role of the accountants has been specially critical. The firm of Price Waterhouse, who vetted the members of the Iranian Consortium in 1954, has for long been the special favourite of the majors. In 1971 they were accountants in the U.S. for four of the majors: Exxon, Socal, Gulf, and Shell. The oil accounting systems are intricate enough thoroughly to confuse any layman. When Price Waterhouse in 1973 produced a survey of the accounting practices in thirty oil companies it revealed such a variety that (in the words of Les Aspin, the Democrat Congressman from Wisconsin) 'no serious students of business economics could adequately compare the relative financial positions of any two of the thirty corporations.'

The tax-avoidance of the companies was their most striking common achievement. These were the figures for their federal tax-payments inside the United States for 1971:

U.S. TAXES PAID BY THE AMERICAN SISTERS

| | 1972 | | 1962–1971 | |
| | Net income before taxes ($ billions) | % paid in U.S. taxes | Net income before taxes ($ billions) | % paid in U.S. taxes |
Company				
Exxon	3·700	6·5	19·653	7·3
Texaco	1·376	1·7	8·702	2·6
Mobil	1·344	1·3	6·388	6·1
Gulf	1·009	1·2	7·856	4·7
Socal	0·941	2·05	5·186	2·7

Source: Multinational Hearings 1974, Part 4, p. 104.

In avoiding taxation the oil lobby was certainly successful, both in America and in Britain—perhaps too successful, for the public indignation in both capitals when the details came to light was furious. But in the field of foreign policy, as the events of the next chapters will show, their success was much less marked—most particularly on the Arab–Israel question. In Britain it was true, the oil companies were well pleased by the emergence in 1970 of the 'Harrogate Speech', in which the Conservative Foreign Secretary, Sir Alec Douglas-Home, moved away from total commitment to Israel. But the speech had originated inside the Foreign Office— where it had been lying around in draft for a year before—rather than in the boardrooms of Shell or BP. And in Washington the oil companies, while trying hard to change U.S. foreign policy, were faced with a heavy disappointment; most of all after the inauguration of President Nixon in 1969.

Nixon had always been regarded as a likely friend of the oil companies, with his Californian base and his Texan allies. The 'slush fund' of $18,000, whose exposure in 1956 nearly wrecked his vice-presidential campaign, included much oil money, including some from Herbert Hoover Junior. In the 1968 campaign he again promised to defend the oil depletion allowance, and was once more heavily backed by big oil, including $60,000 from Robert Anderson of Arco and a disclosed contribution of $215,000 from the Mellons. But soon after his election, Nixon agreed to the reduction of the depletion allowance from 27½ to 22 percent (a change which affected the smaller companies rather than the majors, whose real tax advantage came from the foreign tax credits). In foreign policy, Nixon was no more obliging. After his election a delegation of oilmen called on him, to explain the need to placate the Arabs. This included, Rawleigh Warner of Mobil, Robert Anderson of Arco, John McCloy and David Rockefeller. Nixon, who was accompanied by Henry Kissinger, was very sympathetic, and explained that one group to which he was *not* indebted for his election was the Jewish vote. But a few days later, as Warner later complained, United States policy was back where it had been before. The lobby, then and later, was singularly ineffective on Middle East policy: 'we could always get a hearing, but we felt we might just as well be talking to that wall.'[11]

But the companies continued to contribute generously to the Republican Party, and President Nixon's fundraisers, Maurice Stans and Herbert Kalmbach, leaned heavily on them to help finance the notorious 1972 campaign. Four of the sisters contributed substantially, mostly through individuals. Officials of Exxon gave $217,747 led by the chairman, Ken Jamieson ($2,500), the president Jim Garvin ($3,200) and the head of their Greek affiliate, Thomas Pappas ('the Greek bearing gifts') ($101,672);

while the Rockefeller family gave $268,000. Socal gave $163,000 led by their chairman, Otto Miller ($50,000) and including $12,000 from John McCone. Mobil gave only $4,300, and Texaco (whether through caution or meanness) apparently gave nothing. By far the biggest contributor was Gulf whose offerings included a million dollars given clandestinely by Richard Mellon Scaife, a major Gulf shareholder with his own political ambitions; and at least $100,000 which was produced through the Bahamas subsidiary of Gulf by the chief lobbyist of the company, Claude Wild. The eventual discovery of these illegal gifts, and of others, was to bring back all the old public suspicions of the corruptions of oil money.[12]

But these generous gifts, as we will soon see, did the oilmen little good on the central question of foreign policy in the Middle East. When they came to seek serious help from the administration, whether over Libyan oil, or over Saudi Arabia, the oil money could not offset the old distrust of the companies, or the reluctance of governments to involve themselves.

The global scope of the oil money, however, was not to emerge until 1975, when the Securities and Exchange Commission began investigating political contributions. In April 1975 Gulf were eventually compelled to admit, in their 1975 proxy statement, that between 1960 and 1973 'approximately $10·3 million of corporate funds were used in the United States and abroad for such purposes, some of which may be considered unlawful'. Soon a succession of countries—Venezuela, Bolivia, Peru, Ecuador—demanded to know whether their politicians had been bribed, and Peru even expropriated Gulf's properties. Eventually the chairman of Gulf, Bob Dorsey, had to confess to having paid bribes of $4 million from 1966 onwards to the ruling party in South Korea; and to having given another $350,000, together with a helicopter, to the late General Barrientos in Bolivia. The limelight then shifted to Exxon, whose chairman, Ken Jamieson, had to admit in May 1975 that his company had made political contributions in Canada and Italy; and a new uproar ensued.

What was disturbing was not just the huge bribes themselves; but the fact that they could for so long, and so effectively, have been buried within the company accounts. The ability of a giant corporation, in spite of its auditors, thousands of shareholders and elaborate controls, to conceal such huge sums through underground routes, spotlighted the fact that the big oil companies were, in both the technical and general sense, unaccountable.

NOTES

1. See Edith Penrose: *The International Petroleum Industry*, London, 1968, p. 78.

2. Penrose: p. 146.

3. J. E. Hartshorn: *Oil Companies and Governments*, London, 1967, pp. 375–388.

4. Penrose: p. 263.

5. A. J. Pinelo: *The Multinational Corporation as a Force in Latin American Politics*, New York, 1973, p. 18.

6. The Greeks had traditionally been the diplomats in the old Ottoman Empire, dealing with Arabs on behalf of their Turkish masters.

7. Ironically, it was a Labour government, in 1966, which allowed the government's holding to fall to $48\frac{1}{2}$ percent, thus theoretically losing control; but the Bank of England in 1975 acquired another 20 per cent holding after the crash of Burmah Oil.

8. Speech to API: Houston, 1972.

9. *The American Oil Industry*: the Marine Engineers Beneficial Association, New York, 1973, p. 120.

10. See *The American Oil Industry*, 1973, p. 106.

11. Rawleigh Warner: interview with author, September 1974.

12. See Congressman Aspin's figures in *Congressional Record*, January 23–24, 1974. Also testimony of Claude C. Wild to the Ervin Committee, November 14, 1973.

Libyan Ultimatum

It seems that it is only in the United States that an almost masochistic attack on the position of its own oil companies persists.
John J. McCloy, 1974

It was not in the established oil countries of the Persian Gulf that the sisters faced their first critical showdown, but in Libya, the upstart oil producer on the edge of the Arab world in North Africa. For Libya broke up the ranks on both sides. It had let in the independents to challenge the sisters; and it was aloof from the cautious attitudes of the rest of OPEC. It was the outsider at both ends and by ignoring the rules it changed them.

Since the Libyan oil began to flow in the early 'sixties, it had a fatal fascination for the West, particularly Europe. By 1969 Libya was supplying a quarter of Western Europe's oil. It was of high quality, with little sulphur, which became more important as the West worried more about pollution: and it was very close to Europe, on the right side of the Suez Canal. That became more significant after the Canal was closed in the 1967 war, and still more so after May 1970 when the 'Tapline' from Saudi Arabia was again breached in Syria, and then carefully not repaired. The closeness of Libyan oil was now still more desirable, and there was no alternative so attractive: Nigeria, two thousand miles further south, was being rapidly developed as a 'safe' new source of oil, but by the middle of 1970 Nigeria was being rent by the Biafra War, and supplies had stopped. Libyan oil was not only the closest, but the cheapest, for the companies made no special allowance to Libya for the cheap transport. Exxon argued that if they paid more because the Suez Canal was closed, they would not be able to reduce the price when it was open again. But many oil experts reckoned that the Libyans were being screwed, and that it was only a matter of time before they realised it.

The rush of Libyan oil, like all sudden oil bonanzas, brought with it great dangers to the big companies. In the first place it threatened, as we have seen, to disturb their delicate balancing act, and to cause bitter resentment with the older producers, particularly Iran and Saudi Arabia. The problem was well put by the *International Petroleum Encyclopaedia*, for 1970 (p. 36):

> It's not hard to see why, as increasing amounts of African oil threaten to grab off even larger chunks of their prime market target (Europe), the Mideast nations become upset. Their very way of life is being threatened—a way of life they are just becoming used to and one which they don't want to lose. The interests of the producing and the consuming countries are at once the same and exactly the opposite. The role the large international oil companies play as a buffer element between the two is essential to both. It's a role that, if eliminated would throw the two forces face to face and spell disaster for the entire industry. This then is the reason for the 'three-party' system which is of benefit to all.

The dependence on Libyan oil was also more directly dangerous to Europe, for the Libyans could threaten to cut it off to extract better terms; and some experts in the State Department in the late 'sixties were seriously concerned. There was even a proposal for consumer governments to collaborate in an international treaty to safeguard oil supplies. But the governments were at odds, and the companies were complacent, too busy making money out of Libya.

Libya had also become a bitter battleground between the majors and the independents. From the beginning, as we have seen, the Libyans were determined to bring in the outsiders, to speed up exploration. 'We wanted to discover oil quickly', explained the former Petroleum Minister, Fuad Kabazi: 'this is why we preferred independents in the first stage, because they had very little interests in the Eastern Hemisphere outside Libya.'[1] It was true that Exxon once again led the field, producing 750,000 barrels a day from Libya by 1970; but the independents were producing half the oil from Libya, and they had no interest in restricting production or in playing the balancing game. Exxon and the other sisters were alarmed by their reckless expansion, and exasperated because the independents, as they saw it, enjoyed a tax advantage: for the independents paid taxes on the basis of the market price of the oil they sold, while the majors had to pay on the higher posted price. They saw their chance to damage the independents by using OPEC against them—an ingenious but dangerous game. After OPEC in 1963 had demanded that royalties be included in their 'expenses', the majors offered the same terms to Libya (who had just joined OPEC) provided all companies were taxed in the same way. The independents fought back bitterly, arguing that they were not involved with OPEC, and could not afford higher taxes. But the new Petroleum Law was duly passed in 1965, and the independents never forgave the majors.

Then, in the round of concessions in 1966, there arrived in Libya the most wily and independent of all the independents; the unique phenomenon of Dr. Armand Hammer. This extraordinary old walnut of a man had a combination of imagination and ruthlessness that made him in some ways more disrupting to the sisters than Getty or Mattei; and his whole career had been based on defying convention. He had made his first fortune in Russia after the Revolution and built up a fabulous collection of Tsarist treasures, the first of several art collections. His first experience of oil was in trying to acquire Russian oil for Germany in the 'twenties, when Exxon and Shell were battling for 'stolen oil',[2] and he was thus no great respecter of the sisters. He had come back into oil in 1956, by buying up a sleepy West Coast company called Occidental (or 'Oxy'), and he was now determined to join in the Libyan bonanza.

He made a bid of exceptional generosity, offering an agricultural development project and a joint ammonia-plant; and he wrapped up his bid in red-green-and-black ribbon, the Libyan colours. He also—according to evidence in a subsequent lawsuit—recruited a team of entrepreneurs to help him get the bid. He got his concession, and was soon sensationally successful in his discoveries becoming for a short time the biggest producer in Libya. The advent of Hammer, whose oil was soon flooding across Europe, still further exasperated the majors, particularly Exxon. As the American Ambassador, David Newsome, put it later, with marvellous understatement, 'I think it is safe to say that the advent of Occidental on the scene was not warmly welcomed by all of the other companies'.[3] It was not just that Hammer was adding to the glut; Exxon also knew that Hammer was far more vulnerable to pressure from the Libyans. For him, Libyan oil was his life blood.

So long as King Idris was in power in Libya with his corrupt regime, the oil companies were not seriously threatened. The King complained about the low price of oil, but the warning of Mossadeq was still in the background. Everything was changed on

September 1, 1969, when Idris was deposed by a band of young
army officers, led by Colonel Muamer Qadaffi. They were determined to use oil as an ideological weapon against Israel and to
make the West pay for it. They knew the workings of the oil
business: the first prime minister, Dr. Suleiman Maghrabi, had
taken his doctorate at George Washington University, had worked
briefly as a lawyer for Exxon, and was later jailed in 1967 for
organising an oil-workers' strike. The new government had no
doubt that Libya had been cheated by the oil companies. The
'wild men of Libya' saw oil in the simplest terms, with none of the
sophistication of the Shah or Yamani, but with a directness which
was to dispel the mystique of the sisters, and revive the whole
confidence of OPEC.

Colonel Qadaffi quickly confronted the oil companies; he told
the twenty-one companies that unless they agreed to raise prices
he would take unilateral action. To show they meant business, the
new regime soon made contact with Moscow, to discuss eastern
markets for their oil; and they also began talking to the oil companies—separately, picking them off one by one. They began by
talking to Exxon and Oxy. They demanded an extra forty cents a
barrel, which was not exorbitant in view of the quality and
accessibility of Libyan oil compared to the Persian Gulf. And they
received some support from an unlikely quarter, the State Department.

There the oil expert was Jim Akins, a forthright Arabist who
was increasingly worried by the prospect of an energy crisis, and
will play an important part in the subsequent story. He was a
formidable advocate, a tall erect Quaker with uncompromising
principles; but he was regarded by many diplomats as being too
committed an Arabophile. He was now convinced that the companies must try to come to terms with the Libyan revolutionaries;
he talked to the sisters, including Exxon, Texaco and Mobil, and
also, through the British Embassy, with BP, advising them all that
the Libyan demands were fair.[4] But the big companies, led by
Exxon, were adamant: they would not pay more than five cents
extra. Many oilmen reckoned that Exxon were being excessively
hard-nosed, and that settling now would save trouble later. But
Exxon were in a tooth-and-claw mood, and their major interests
were anyway outside Libya.

Qadaffi lost no time in reprisals. He struck not at Exxon but
at the newcomer Oxy, which he well knew was totally dependent

on Libya. In May and June 1970 he ordered Oxy to cut back production from 680,000 to 500,000 barrels a day. Officially it was for reasons of conservation—the old pretext for cutbacks—and certainly Oxy was extracting oil at a rate which many oilmen thought was harming the field. But the cutbacks were obviously meant to force Oxy to pay more, and they soon had their effect. Oxy, with their own refineries in Europe waiting for Libyan oil, could not get the oil anywhere else; and Oxy shares were a 'hot stock' which depended on Libya. By July, Armand Hammer was desperate, and he went to see Ken Jamieson, the new chief executive of Exxon. It was a historic meeting between the two opposite sides of the oil business. Hammer told Jamieson he could not withstand the Libyan pressure unless he had an alternative source of crude oil. Could Exxon help? Jamieson was more conciliatory towards the independents than his predecessor, Haider: but he still did not trust them. He was prepared to promise Hammer replacement, at the normal price for third parties. But Hammer wanted oil at close to cost, and Jamieson could not agree.[5] Thus Exxon rejected the first opportunity for a common front; it was some time before Jamieson realised the extent of the common danger.

Dr. Hammer was thus left on his own. He himself negotiated with the revolutionaries through August, flying back from the sweltering heat each night to Paris in his private plane. But he had no real leverage, for both sides knew that he could not do without the oil. Finally he agreed to pay thirty cents more, going up a further two cents a year for five years, and to raise the tax rate from 50 percent to 58 percent.

Other company men were appalled: as a Shell man put it, 'from that point on, it was either a retreat or a rout.' Two weeks later the Libyans made a similar deal with the Oasis consortium, whose shareholders included three independents, Continental, Marathon and Amerada-Hess. But another of the shareholders was Shell; and Shell with their huge interests elsewhere, were in no mood to give in. Their chairman since 1967 was an outspoken aristocrat, Sir David Barran, who was a figure of stature in the oil world: he was precisely articulate, wore a monocle and had the style of a cultivated country squire. He believed that, even though Shell depended heavily on Libyan oil, they must not give in; for that would risk undermining 'the whole nexus of relationships between producing governments, oil company and

consumer'. He believed that Shell should 'at least try to stem the
avalanche'.[6] Shell therefore refused to sign the agreement. A
week later, on September 25, all its share of Oasis production
(150,000 barrels a day) was stopped.

In New York, the American companies were now thoroughly
alarmed, and they turned again to the master-lawyer who had
served as their legal adviser off and on for the past decade: John
Jay McCloy. After he had told President Kennedy about the
dangers of OPEC in 1960, McCloy had warned successive
attorney-generals that the oil companies might eventually have
to act together; when in late 1970 the seven sisters saw OPEC
threatening them with a succession of escalating demands, they
looked naturally to McCloy. He was now seventy-five, with a
unique experience of governments and oil policy, working from
his own law firm at the top of the Chase Manhattan skyscraper in
Wall Street. He represented not only all seven of the sisters, but
nearly all the biggest independents, too. Administrations, trust-
busters, and chief executives came and went, but McCloy carried
on, as the memory of the industry, and its link with each government.

Three days after Oxy had given in to the Libyan demands,
McCloy went to Washington with the heads of the oil companies
to talk to the State Department. They saw the Secretary of State,
William Rogers, Under-Secretary Alexis Johnson and Jim Akins
the oil expert. They agreed that the position was serious, but
reached no decision. Two weeks later the heads of the two British
oil companies, both involved in Libya, came over to New York:
Sir David Barran of Shell and Sir Eric Drake of BP. They
lunched together with the British Foreign Secretary, Sir Alec
Douglas-Home, who was attending the United Nations, and
Barran explained to him the gravity of the crisis and the danger of
an avalanche.[7] They told Sir Alec that they thought the oil companies should try to hold out, even at the risk of losing their
Libyan concessions. They reckoned that without Libyan oil the
companies should be able to supply Europe with 85 to 90 percent
of its needs for at least six months, probably without need of
rationing; by that time there would probably be either a settlement, or further alternative sources. Sir Alec was sympathetic and
said he would consult his European counterparts at the U.N.: but
having done so later, he reported a noticeable lack of preparedness
among them to risk any cutting-off of Europe's oil. The prospect

of a common stand was thus already dim; for without the governments behind them, the companies knew their position was perilous. If the companies were nationalised, the Europeans might well buy the 'hot oil' direct from the Libyans, and thus undermine them: with the growth of the independents and national companies the majors were no longer in the position to enforce a boycott, as they had been in Iran twenty years earlier.

Just after their lunch, Sir David and Sir Eric flew on to Washington to attend a further meeting at the State Department, headed by Alexis Johnson, and attended by McCloy and the heads of the American sisters. The British pair found it an astonishing meeting. For the first hour, the oilmen were lectured about the problem of Jordan and the Palestinians, with the implication that a Middle East settlement would also settle the oil problem. At last Barran was able to put his case, that the seven must stand together over Libya: Shell's experience in Libya gave his argument, as he put it, 'a rather specially keen cutting edge',[8] and he even suggested that the American companies should dare the Libyans to nationalise. He was supported first by Drake of BP, and then the the chairman of Mobil, Rawleigh Warner; and by Mobil's president, Bill Tavoulareas, now making his debut in oil diplomacy.

But is was now clear that the 'terrible twins' among the seven, Socal and Texaco, were in no mood for a showdown: they had a joint concession in Libya which was threatened, and they changed quickly from their customary hawkish position to a dovish one. As for Jamieson of Exxon, he sounded statesmanlike, holding the balance between the doves and the hawks, without giving an opinion. Some of the American oilmen seemed infected by Sir David's boldness, and by his assumption—so contrary to anti-trust principles—that companies and governments should work closely together. But Jim Akins, for the State Department, was against provoking the Libyans. Akins, according to one of the British oilmen 'was hypnotised by the Saudi Arabians. He said that there was no question of Saudi Arabia following Libya. I said you must be joking and nearly walked out.'

Sir David flew back to New York in the Exxon plane, with Jamieson and McCloy, and told them his suspicions of Socal and Texaco; he thought the game was up. Shell stuck it for for a few weeks, but soon Socal and Texaco *did* cave in, on terms very similar to Oxy's. Their colleagues suspected that they were not averse to putting up the price to undermine the independents:

it was anyway Aramco they cared about, far more than Libya. The surrender by the two sisters was more significant than Hammer's; and the others soon followed. Sir David and his board decided that continued resistance, and consequent isolation, became pointless'. The Libyans had decisively won the first round, and the companies were in visible disarray.

The demands for higher prices were now rapidly spread beyond Libya, with the 'leapfrog' effect which the oilmen had always dreaded. The frog leapt to Iraq, Algeria, Kuwait and Iran, which all quickly claimed an increased tax rate of fifty-five percent. And in December 9, 1970 the members of OPEC met in Caracas in a mood of new militancy. As one OPEC official later put it: 'The Libyan success was an embarrassment to other OPEC countries. It rendered further silence almost impossible.'[9] The more radical members saw the Libyan tactics with the technique of the cut-offs as preparing the way to new victories.

At the same time there were now the first signs since the formation of OPEC of a world shortage of oil. OPEC's report saw 1970 as a turning-point, with the buyer's market turning to a seller's market. What the moderate leaders of OPEC had failed to achieve in ten years, the tactics of the wild revolutionaries of Libya were apparently achieving in a few months. At Caracas it was decided that there was now a 'change of circumstances' as a result of the market situation, which justified revising agreements: and a new resolution was adopted, declaring 55 percent as the minimum tax rate on profits, advocating higher posted prices, and eliminating discounts for companies. OPEC also resolved on 'concerted and simultaneous action', and proposed a new round of negotiations with the companies in Teheran in the New Year. But even these proposals were not militant enough for the Libyans, and soon after Caracas, the frog leapt again: Libya demanded another fifty cents a barrel, with retroactive claims and an extra twenty-five cents for 'reinvestment requirement'.

The avalanche was now rolling. But in the meantime some of the oilmen were trying again to form a barrier against it. Dr. Hammer, realising that his own capitulation had begun the retreat, was in touch with Sir David, to explain his predicament (it was Hammer's habit to ring up his fellow oilmen from California, unaware that he was waking them up in the middle of the night). At the end of 1970 he asked Sir David whether Shell would be able to help out the independents with a 'safety-net' in the next showdown (as he had asked Jamieson six months before). Sir David wasn't sure about the legal aspects, but promised that if the rest of the industry agreed to a plan, Shell would join in. After observing the escalating demands from Caracas and Tripoli, Sir David was convinced that this was the last chance for bold action. He wrote a 'New Year Letter' to all the oil companies concerned with Libya, proposing a meeting to discuss a joint policy and suggesting they should all decline to deal with the producers except on a total, global basis.

As a result on January 11, 1971, the representatives of twenty-three oil companies assembled in New York in the lush offices of John McCloy. All seven sisters were represented, as were the leading independents, CFP, the Japanese Arabian Oil Company, the Belgian Petrofina and the German Elverath. The immediate problem was the anti-trust laws, on which McCloy was expert, and he was soon able to show his influence. In his anteroom, decorated with signed photographs of every past President since Roosevelt, were waiting two men from Washington. One was Jim Akins from the State Department, the other was Dudley Chapman from the anti-trust division: and McCloy explained that they would inspect any agreement and prepare for a clearance from the Justice Department. Over the next three days, the company executives continued meeting in New York—in McCloy's office, in the Mobil skyscraper in 42nd Street, in the heavy palazzo of the University Club (which since the first Rockefeller joined it, has been a haven for oilmen)—while the government men waited to refer drafts of their agreement back to Washington.

It was the collaboration between government and companies for which McCloy had been holding himself in readiness for ten years. His justification was persuasive, for the oil companies were now up against a more formidable potential cartel than themselves. As he put it: 'the idea that you can't confront a highly organised cartel of sovereign states is rather silly.'[10] But the collaboration marked a total reversal of the ostensible principles of anti-trust. Now (as in Iran in 1954) anti-trust appeared as a luxury which could be dispensed with in time of crisis. And more important, the collaboration assumed that OPEC could only be dealt with by confrontation. The separation of interests which had begun in 1960 had now developed into a full-scale clash between two armies.

THE 'POISONED LETTER'

The oilmen now had two aims before the Teheran meeting: the first was to write a joint letter to OPEC, proclaiming their common front. The second was to sign a 'safety net agreement' to help each other out, in case they were again picked out one by one. The letter was finally approved in the middle of the night in the Mobil headquarters, and signed by the representatives of twenty-three companies taking part. The letter marked a re- markable reversal from the companies' first attitude to OPEC ten years ago. Then, they were refusing to recognise its existence; now, they were insisting that it must be effective and binding on all its members. In confronting each other, the two cartels were building each other up.

The letter began mildly enough with the words: 'We wish to place before OPEC and its member countries the following proposal', but it came quickly to the point: 'we have concluded that we cannot further negotiate the development of claims by member countries of OPEC on any other basis than one which reaches a settlement simultaneously with all producing govern- ments.'

This accomplished, the companies then discussed the terms of the safety net, and prepared the 'Libyan Producers' agreement'— a document which was kept secret for the following three years.[11] Each party promised not to make any agreement with the Libyan government without the assent of the others; and if the Libyan government ordered one to cut back, all the others would share the cutback in specified proportions. John McCloy had obtained Washington's approval in the form of a temporary 'business review letter' from the head of the anti-trust division, Richard McLaren. It was not (McCloy insisted) the same as a waiver; it was 'pretty left-handed'. But in effect it guaranteed that the Justice Department would not interfere with the collaboration. For McLaren, who at this time appeared to be embarking on an anti-trust crusade, notably against ITT, this was a striking con- cession to the oil companies.

McCloy had also taken a new diplomatic initiative. On January 15 he again went to Washington together with several heads of the oil companies to visit the Secretary of State, William Rogers, on the eighth floor of the State Department. The other diplomats present included Alexis Johnson, Jim Akins and Jack Irwin.[12]

a former lawyer with Sullivan and Cromwell, who had just joined the Department. McCloy explained that, in addition to the message to OPEC and the safety-net agreement, 'it would be wise if the government could enter into this thing and get the heads of the countries involved to moderate their demands'.

The oilmen suggested that the State Department might send a diplomat to the Middle East. Secretary Rogers asked who, and someone suggested Jack Irwin.[13] Rogers discussed it with President Nixon the same day, who then gave a personal message through Irwin to the Shah of Iran, the King of Saudi Arabia and the Sheikh of Kuwait, expressing his interest in the Persian Gulf and his concern about oil supplies. The very next day Irwin flew out to the Middle East: 'My mission,' as he explained it, 'was to stress to the leaders of the countries the concern the United States would feel if oil production were cut or halted.'

In the meantime the companies' letter to OPEC members had rapid repercussions. In Libya, the representatives of the two most threatened independents, Oxy and Bunker Hunt, had stormy interviews with the Oil Minister, who alternated threats with enticements to break away from the other oil companies, and harangued them about the 'poisoned letter' to OPEC. But they stood firm. The Bunker Hunt representative, Henry Schuler, was a tough-minded young Princetonian who was convinced that the companies could out-stare the countries, and his boss Bunker was right behind him. Schuler believed that the Libyan bravado was bluff.[14]

In Iran too the reaction was explosive. The finance minister, Dr Amouzegar, was now in charge of oil policy, a sophisticated and cosmopolitan diplomat, articulate in several languages, with a disarming humour: he had studied hydraulics at Washington University, and had married a German wife. Dr Amouzegar immediately protested. A single negotiation, he warned, would be disastrous for the oil companies, for the other OPEC countries could not stop the Libyans making 'crazy demands', and would then be committed to following them. The Libyans would not join the Gulf States, and the Gulf States could not wait. The joint approach was a 'dirty trick' which could lead to the oil from the whole Persian Gulf being shut down.

Two days later Jack Irwin arrived in Iran, at the beginning of his tour, and went to see the Shah, accompanied by the American

Ambassador, Douglas MacArthur (the son of the general). The Shah reiterated Amouzegar's arguments. The American diplomats were persuaded with remarkable speed, in view of the strong language of the letter to OPEC which they were supposedly supporting: McCloy suspected that MacArthur was suffering from the ambassadors' ailment of 'localitis', becoming more Persian than the Persians. MacArthur quickly recommended to Secretary Rogers in Washington that the companies should have two separate negotiations.

'BEWARE THE OIL KINGS'

The day after this setback for the companies, the two sides gathered in Teheran to begin bargaining. OPEC had appointed a formidable trio, all educated in America: Dr. Amouzegar, for Iran; Sheikh Zaki Yamani, the Saudi oil minister; and Saadoun Hammadi, the austere chairman of the Iraq National Oil Company, who had taken his Ph.D. in agricultural economics at Wisconsin University. The companies, on their side, had chosen two chief negotiators: the first was George Piercy, who had recently taken over from Howard Page as the director of Exxon concerned with the Middle East, and who from now on was probably the most important single figure in oil diplomacy. He had shown himself adaptable and conciliatory in dealing with Middle East politics, but he lacked Page's intuitive understanding of the Arab attitudes, and particularly of Yamani; and he was still new at the job. The other was Lord Strathalmond, the son of the former chairman of BP, who had just succeeded to his father's title and was now a managing-director of BP.[15] The second Baron was a much more genial man than his dour father. He had begun as a lawyer, and had mde his name in BP by arranging for the tax paid by BP in Kuwait to be exempted from British taxation. He kept racehorses, enjoyed night-life, had a house in Tobago and an American wife. He was on excellent terms with both sides; he enjoyed arguing over gin and caviar with Amouzegar till two in the morning, and he got on very well with 'Groucho', as he called the Kuwaiti oil minister, Atiqi. But he was not the intellectual equal of his opponents, and he freely admitted that Amouzegar had a far finer brain.[16] Some of the Iranians were understandably confused by his arrival, since they associated his name with their bitter negotiations fifteen years before.

They arrived in Teheran at short notice in some confusion, of which the details, as recorded in the cables, have recently been painfully revealed.[17] Piercy, having helped to prepare the letter to OPEC, was surprised by Ambassador McArthur's recommendation, and he cabled back to New York that it would violate the companies' letter.[18] He soon realised that the diplomats had undermined the companies' strategy, and the next day the two negotiators cabled to headquarters: 'It was perfectly clear that Dr. Amouzegar believes he and His Imperial Majesty have convinced American Government in recent discussions of the correctness of their position on a Gulf negotiation coming first, with the result that our negotiating stand on the procedure to be adopted is by no means an easy one.'

The State Department had already hindered rather than helped the companies ('We weren't too impressed' said McCloy), who now had the worst of two worlds. Their common front had first provoked OPEC and then broken in two. In the muddle the Shah was soon able to drive a wedge between the companies and the government, and gave the oil companies a two-day deadline to abandon their global approach. Piercy tried to enlist the support of Yamani, explaining that they had a mutual problem over Libya's preposterous demands, which were not in anyone's interests. Yamani was sympathetic, but said that OPEC could not control the Libyans, and he warned Piercy significantly: 'George, you know the supply situation better than I. You know you cannot take a shutdown.'

Yamani also confirmed to Piercy a rumour that he had already heard: that at the OPEC conference at Caracas, six weeks before, there had been a plan to enforce a world oil embargo to strengthen their demands, which (said Yamani) had the blessing of both the Shah and King Feisal. Piercy said the companies were astounded, and warned Yamani about the effects of an embargo, and 'what it will do to the prestige of these producing countries'. But Yamani explained—as he was to explain many times later—'I don't think you realise the problem in OPEC. I must go along.'[19]

Strathalmond and Piercy cabled in bewilderment now hoping that OPEC, their old enemy, would hold together: 'It is not easy to advise what should be done. If we commence with Gulf negotiations, we must have very firm assurances that stupidities in the Mediterranean will not be reflected here. On the other hand, if we stick firm on the global approach, we cannot but think . . . that there will be a complete muddle for many months to come. Somehow we feel the former will in the end be inevitable.'

In New York, in the meantime, 'the Chiefs' of the companies were assembling daily, usually in the Mobil offices by Grand Central Station, still hoping that a common front could hold together. With them periodically was John McCloy, now playing a double role; being both the companies' lawyer and the agent of the Justice Department to 'monitor' the discussions to make sure that they did not connive beyond their agreement.

At the same time in London senior executives were assembling for the new committee formed as a result of the joint agreement, called The London Policy Group, or LPG. They met at the new skyscraper of BP, Britannic House, and American oilmen were struck by the smooth organisation; it was like a peace conference of diplomats. Through the thickly carpeted foyer with its models of tankers, past an anteroom full of refreshments and handouts, they entered the great BP conference room in the basement with a row of fifty black leather chairs facing the tables with another row of chairs behind for advisers.

It was a gathering of oilmen unique in the history of the industry.[20] At the head was the veteran Joe Addison, the chairman, just retiring as head of the Iranian Oil Participants, with beside him Bill Jackson from McCloy's office. On the right were the American companies, headed by Exxon, Mobil and Texaco, going down to Occidental at the end. On the left was Shell, followed by BP, Marathon, Gelsenburg, Hispanoil and CFP. BP and Shell, untroubled by past memories of anti-trust, dominated the meetings with their quiet assurance and copious information; when for instance the Americans were worried that OPEC might be planning changes in their currency demands, Shell were able to reassure them that their men had been watching the OPEC offices in Vienna: no monetary experts had come in, and no OPEC economists had gone out. There were no diplomats present, but Jim Akins was in London and kept himself informed.

The assembly included some of the best brains in the oil business—Tavoulareas of Mobil, far more incisive than his American colleagues; Jean Duroc Danner, the brilliant director of CFP; David Steel, the lawyer who was heir-apparent of BP. George Piercy of Exxon arrived with a retinue of experts worthy of a summit conference. Laurence Folmar of Texaco lived up to his company's dogged reputation. The frictions between the majors, and between them and the independents, were very soon evident, and the independents quickly showed their anxiety. 'Smokey' Shafer of Continental veered between industrial statesmanship and pre-occupation with his company's dependence on Libya. George Williamson, the brash young representative of Oxy, faithfully re-flected the toughness of Dr. Hammer. It was very clear that the companies did not altogether trust each other, with some reason. In later London meetings the mood was not improved by the dis-covery that the Libyan ministers, Jalloud and Mabruk, appeared to have full knowledge of the terms of reference of the company negotiators. Someone in the room, it seemed, had been leaking.

For three weeks confused cables passed between the four bases (Teheran, London, New York, later Tripoli) and executives flew across the Atlantic and the Mediterranean. But the cross-purposes became still more apparent. At the first London meeting, the oil companies had already backed down from their global approach without actually admitting it.[21] The independents, led by Schuler of Bunker Hunt, protested at the climb-down, but did not use their veto. The meeting concluded that 'we would not exclude that separate (but necessarily connected) discussions could be held initially with groups comprising fewer than all OPEC members'. But the phrase 'necessarily connected', as Schuler complained, was a moving target; 'it kept changing its significance as we encoun-tered one round of resistance after another.'[22]

The London Policy Group now drafted a new letter to OPEC, including the possibility of separate-but-necessarily-connected discussions, and cabled it to Strathalmond and Piercy, in Teheran. Dr. Amouzegar described it as a 'poor lawyers effort', and Strathalmond was inclined to agree. Amouzegar insisted that the companies must start negotiating immediately with the Gulf countries.

Piercy flew back to New York, Strathalmond flew to London, and the two sides exchanged intercontinental missiles. In Teheran the Shah, now in his element, gave a press conference for two-and-a-half hours attacking the oil companies for enlisting the support of their governments — 'a precise example of what is called economic imperialism'. He threatened to remove the companies altogether: 'the conditions of the year 1951 do not exist anymore. No-one in Iran is cuddled under a blanket or has shut himself in a barricaded room.' But he also gave his fellow-members of OPEC a pregnant warning: 'if the oil producing countries suffer even the smallest defeat, that would be the end of OPEC. And then, nations will not dare to gather together and to rise against these giants.' He recalled how he had once been told: 'Beware the Oil Kings and what they might do.' But now, he made clear, the companies had lost their old power.

In New York, McCloy was worried by the apparent retreat. He took a tougher line than any of his clients, and while they explained to him that it was their money that was at stake, he insisted that if the producing governments held together, the companies must hold together too.[23] Moreover the anti-trust chief, McLaren, warned McCloy that his anti-trust clearance only related to the original message to OPEC, insisting on a single negotiation. In the meantime the U.S. government conceived a special meeting of OECD in Paris where the consuming nations made clear that they were in no mood to resist higher prices.

But the oilmen were now alarmed that the Gulf countries might take unilateral action, and they clung to the hope of 'separate-but-necessarily-connected'. They took refuge in a new notion, that the Gulf settlement could be a 'hinge' for a settlement in Libya, with no leap-frogging between them. So they agreed to split into two teams: one to stay in Teheran, led by Strathalmond, to treat with the Gulf states; the other, led by George Piercy, to go to Tripoli to negotiate with the Mediterranean states, led by Libya.

In Teheran on January 28, Strathalmond at last began the formal negotiations; the Gulf countries had now set a five-day deadline. Strathalmond proposed the companies offer, including an increase in the posted price of $0·15 per barrel, and allowances for inflation: Dr. Amouzegar demanded an extra $0·54 a barrel, and a much higher inflation factor. The producing countries threatened a world-wide shutdown and the Libyans refused to negotiate in Tripoli until the Teheran terms were agreed. The team returned to London to consult.

The London Policy Group now discussed enlisting their governments' support; but having consulted with the 'Chiefs' in New York, they decided instead to put their faith in assurances from OPEC. Henry Schuler was horrified by the retreat, and at a dinner on January 30 he argued his case hotly, backed up by his colleague, Norman Rooney, who burst out: 'You're selling us down the river!' Next day he even talked on the telephone to the chiefs in New York, at the invitation of Jamieson of Exxon, and protested that the retreat would destroy the companies' credibility on all sides. But only the German Gelsenburg supported him,[24] and reluctantly he gave in, while putting his opinion on record. As he put it later: 'the OPEC countries were confident of their ability to face-down the oil companies . . . and the companies had reverted to an attitude of narrow self-interest.'

The companies' ineffectual contacts with governments now provided useful ammunition for OPEC. On January 30 the LPG gave details of proposed terms, in a cable sent through the Foreign Office in London to the British Embassy in Teheran, which thus routinely carried the signature of Sir Alec Douglas-Home. But a public relations man left a copy of it on a table, and the text appeared in full in the Teheran morning papers, to the intense embarrassment of the companies. Sir Alec's signature was removed, but Dr. Amouzegar made the most of it in private, commenting that if the censorship were removed, it would become clear that the British government was running the negotiations.

The deadline expired on February 2, and negotiations were broken off. OPEC then held its own conference, ending with a menacing resolution that each country would legislate new terms if the companies did not accept by February 15, and would embargo any country which did not accept the terms. The companies faced a simple choice: settle, or be settled.

But the majors were desperate to avoid unilateral action, or nationalisation. The team returned to Teheran, where the Shah was now less menacing, but still firm: 'all the oil producing countries know that they are cheated,' he told the BBC, 'otherwise you would not have the common front . . . the allpowerful six or seven sisters have got to open their eyes, and see that they're living in 1971, and not in 1948 or 9'.

On St. Valentine's Day, February 14, a day engraved in the memory of the oil industry, the Teheran agreement was signed. It allowed for an extra $0·30 on the posted price, escalating to $0·50 by 1975: a minuscule increase compared to what happened later, but regarded at the time as almost ruinous. The Shah pledged that there would be no leap-frogging, but the agreement specifically excluded any commitment to oil prices in the Mediterranean. The 'hinge' was either broken or, worse still, the door might now swing back again the wrong way.

McCloy insisted that it was a kind of victory, justifying the combined action. He told McLaren in a long letter in July: 'while the cost of settlement was extremely high, the companies by virtue of their common stand were able to resist the joint threats of OPEC . . .' But many of the oilmen believed that the balance had now turned. The Oil Kings were no longer the companies, but the countries.

THE HOUSE OF CARDS

Four days after the Teheran agreement, talks began in Tripoli. The Libyans were determined not to face the companies en bloc. George Piercy of Exxon had arrived three weeks earlier to lead the companies' team, but the Libyans ignored him, and were determined to deal with the companies one by one. The Libyans were backed up by all the Mediterranean oil countries, Algeria, Iraq, and Saudi Arabia, who soon threatened a Mediterranean embargo if the companies did not agree with Libya's terms. Two of the countries, Iraq and Saudi Arabia, were also Gulf producers: the Saudis, who had earlier been friendly to the companies, were now worried by the danger of their pipeline being blown up by Palestinian guerrillas if they did not join the embargo. The enmity to Israel was adding to the unity.

The Libyan team was headed by the fiery Major Abdul Salaam Jalloud, translated by the veteran diplomat, Ali Mabruk. The Libyans, they pointed out, had survived for 5,000 years without oil, and could quite well do without it, but the companies couldn't. They invited the company representatives, one by one, to negotiate, which meant long waiting followed by insults and speechmaking. When the representatives of Socal and Texaco came to see them with their offers, they looked at them and then threw them to the floor: the oilmen meekly picked them up—a symbolic obeisance from the terrible twins. There was now a visible divergence between the majors, whose main interest was outside Libya, and the independents who desperately needed to keep their Libyan oil cheap. Many independents suspected that the majors would not be too worried to see them suffer.

By March 1971 the companies agreed on a figure of $3·30 per barrel for the Libyan posted price—an increase of 76 cents, together with other fringe benefits. The Mediterranean oil ministers insisted on $3·75, and threatened to cut off all supplies. OPEC made a display of solidarity: the Syrian deputy prime minister flew to Tripoli, Nigeria said it would join OPEC, and eventually Yamani of Saudi Arabia flew into Libya.

The Libyan ministers, Jalloud and Mabruk, at last asked the

German representative of Gelsenburg, Enno Schubert, to negotiate on behalf of all the companies, presumably expecting that a
German would feel specially dependent on Libya. Schubert asked
to be joined by the Mobil man, Andrew Ensor, a patient ex-
diplomat with a long experience of oil policies. The two then
spent seven hours with the two Libyans in a marathon haggle,
with Ensor entering into the spirit of the game, until finally they
narrowed down to two cents, in a classic dialogue:

> *Mabruk:* The Major (Jalloud) really appreciates your putting
> your cards on the table.
> *Ensor:* That's where they are, all right. It's most frustrating to
> be only two cents apart—one permanent, one temporary.
> *Jalloud:* Well, I will move to my final final final. I may be killed
> but I will make both cents temporary.
> *Ensor:* [after further pause] Major, you have been extraordinarily
> accommodating. You know our situation. We are already 1½ cents
> beyond what we have. All the same, you have been so helpful that
> we must make one last effort to close the gap. If you can, here and
> now split the difference at $3·29 we will. I can only get fired once.
> If I would have been fired at 1·5 cents, I might as well risk that
> for 2·5 cents.
> *Jalloud:* You have been very frank. I will be, too. I absolutely
> cannot go below $3·30. I told the Revolutionary Council we
> would get $3·45 but I can persuade them it is necessary to go
> below that, but absolutely not below $3·30. You must understand
> this is psychological. $3·29 sounds much lower than $3·30. So, I
> appreciate your offer but I cannot accept it. I would be killed.
> *Ensor:* That is too bad, but I do understand. I must withdraw
> the offer and we remain 2 cents apart on temporary. If you could
> agree to a recess of an hour or so, we would recommend it. We
> shall tell them how hard you have tried to accommodate us and
> we should hope to come back with the two cents.
> *Jalloud:* Very well. To go lower would be prison, at least for
> me . . .

They adjourned for food, then went on till one in the morning.
They agreed on the posted price of $3·30, with premiums bringing
it up to $3·45.

On April 2, six weeks after the Teheran agreement, the Libyan
government signed the five-year Tripoli agreement with the fifteen
companies: agreements with Nigeria, Iraq and Saudi Arabia
(characteristically the last) followed in the subsequent weeks.

The two agreements in Teheran and Tripoli were meant to hold good for five years, until 1976. They survived for two years, but their weakness was apparent from the start. The Shah was furious when he heard the Libyan terms, realising that he, too, could have got more, and he soon obtained an extra premium for port costs; the door was swinging back again already.

The companies had revealed at Teheran and Tripoli their fundamental weakness, that they could not collaborate either with each other, or with their governments. The collapsed common front had underlined their disability. OPEC had called their bluff and found a new confidence: as one Shell director put it 'they were on the pigs' back, and they knew it'. Moreover, there was now a serious possibility of world shortage, or at least a lack of surplus capacity: the projects outside OPEC, in Alaska or the North Sea, were not developing fast enough to satisfy the West's hunger for oil. As Yamani had warned Piercy, the companies could not face a shut-down in one critical country.

The more far-sighted oilmen realised that the agreements were very fragile: that they were, in the phrase of John D. Rockefeller I, 'ropes of sand'. As one delegate put it, Teheran was 'a house of cards', waiting to be blown down by the next high wind: and Walter Levy, the oil consultant in New York, warned that the oil industry was facing a 'hurricane of change'. What the agreements did was to buy time, for both companies and consumer governments to face up to the next crisis. But almost nothing was done. The more public-minded of the sisters, notably Shell and Mobil, did try to warn their governments and their public. But most of them preferred to present themselves as masters of the situation. They were still playing Atlas, with the world on their shoulders; but privately they suspected that it was beyond their control.[25]

NOTES

1. Multinational Report, 1975, p. 98.
2. Armand Hammer: *The Quest of the Romanoff Treasure*, New York, 1936, p. 143. (The title has little connection with the subject-matter.)
3. Multinational Report 1975, p. 99.
4. Multinational Hearings: Part 5, p. 6.
5. Jamieson: interview with author, May 1974.
6. Letter to Senator Church 1974: Multinational Hearings.
7. Barran: interview with author, October 1974.
8. Letter to Senator Church: August 16, 1974.

9. Abdul Amir Kubbah: *OPEC Past and Present*, Vienna, September 1974, p. 54.

10. Interview with author: September 1974.

11. It was only eventually released under strong protest from the State Department: see Multinational Hearings: Part 5, p. 100.

12. Now U.S. Ambassador in Paris.

13. Irwin was under the impression that McCloy suggested him, but McCloy denied it; in any case the idea apparently originated with the oil companies. See Multinational Hearings: Part 5, pp. 147, 155, 263.

14. Multinational Hearings, 1974: Part 5, pp. 75–101.

15. He was to be appointed as a director of Burmah Oil in 1975 to help rescue the company after it had asked for government aid.

16. *BP Shield*, 1974.

17. See McCloy's testimony in Multinational Hearings: Part 5, p. 62–73.

18. See cables reproduced in Multinational Hearings, 1974: Part 6, pp. 60–65.

19. Multinational Hearings: Part 6, p. 71.

20. I am indebted to an eyewitness for the following description.

21. It has never been the intention that the individual negotiations of the several companies with the several governments [said the agenda for January 20 in stately prose] should be carried out to the last detail by a central, and therefore monstrous, overall negotiation (though this was the implication of their letter to OPEC).

22. Multinational Hearings: Part 5, p. 132.

23. Interview with author.

24. Multinational Hearings: Part 5, p. 135.

25. Some company men saw the fault as lying more seriously with Western governments, who were now more visibly unprepared to take responsibility. This is the comment of one of the company negotiators, on reading this chapter:

Consuming governments preferred to wring their hands, to acknowledge their lack of staff qualities for this sort of negotiation, and their desire to 'keep these matters out of politics', to bicker among themselves and to contemplate *sauve qui peut* initiatives against one another. The Italians, French and Japanese on the whole saw the weakness of the companies, and planned separate initiatives vis-à-vis OPEC; the Americans, British, Germans and Dutch preferred to hope that the hurricane would pass without causing much more damage.

But none of them saw what was really needed—solidarity among all of them—to be a practical policy aim. The French were too anti-American, the British too bemused by their debt to Pompidou, the Italians too much governed by Mattei's legacy of dislike for the seven sisters, and the Japanese too scared of the OPEC reaction for any discussion to get started towards what was needed.

All these governments had no conception of the scale of the disaster to which their lack of initiative and solidarity was exposing them. In this climate, the companies had no sound alternatives; they saw themselves as damned if they did, and damned if they didn't. Certainly, to have publicly drawn attention to their own weakness, as some critics now say they should have done, would not only have been against the grain (no one readily acknowledges he is a broken reed); more important, it could only have hastened the debacle. It would have publicly invited OPEC to drive on further and faster.

11

The Crunch

The price of an article is exactly what it will fetch.
                              Marcus Samuel of Shell, 1911.

After the Teheran and Tripoli agreements, for the next two-and-
a-half years, the companies were in trouble on three separate
fronts. First the producers were demanding part-ownership of
the concessions, or 'participation'. Second, there were increasing
signs of an oil shortage. Third, the Arab-Israel situation was again
heading towards conflict. The convergence of the three was to
produce the greatest crisis in the history of world oil.

PARTICIPATION

For the more radical Arabs the whole concept of oil concessions—which had been leased to the companies under quite
different regimes and conditions for very long periods—was increasingly unacceptable. The notion that a producing country was
entitled to part-ownership of the concession—as opposed to merely
taxing it—had been enshrined as long ago as 1920, as we have seen,
in the San Remo agreement over Iraq, and ever since ignored.
In the meantime the angrier radicals had favoured the more extreme course of wholesale nationalisation. But that drastic remedy,
whether in Mexico in 1938 or in Iran in 1951, carried the danger of
excluding the producers from the oil markets. How could producers take over control, and yet remain part of the world market
system? In the late 'sixties they began to concern themselves with
the idea of gradual nationalisation, under the tactful slogan of
'participation'.

The chief advocate of participation was not one of the fiery
extremists from Libya or Iraq, but the man who had been the
special favourite of the oil companies, a graduate from Harvard,
from the most conservative country of all. Sheikh Zaki Yamani
of Saudi Arabia was to be the dominant figure in the next stage
of the conflict. He had gained great confidence since he had first
become King Feisal's oil minister in 1962. His old oil company

friends, who used to regard him as a clever young protegé, now found themselves being kept waiting outside his huge office in Riyadh, or pursuing him as he flitted between capitals, exercising his growing power with obvious relish. He was quicker than most of them; with his sharp eyes, his Mephistophelean beard and his debonair style, he was like an unbottled genie from Arabia, impatient to take over from his master.

Yamani had been a director of Aramco since 1962, but his apparent involvement only made him more aware of his exclusion from the crucial decisions. As he put his case later, 'I want to know what is going on and I want to have a say in it. And that is what frightens them most,'[1] Yamani realised the dangers of abrupt nationalisation, and the need to maintain the orderly system of the oil companies; and in this he was at loggerheads with his predecessor Tariki, who was now advising other oil countries, and firmly advocating nationalisation. Yamani believed that the oil producers should gradually edge their way into the intricate world system, without disrupting it. The producing countries could take over a share of the oil, and could thus build up their own national oil companies—Petromin in Saudi Arabia, KNPC in Kuwait, NIOC in Iran. In other words, OPEC would not beat the cartel, but join it.

Yamani had been interested in some form of participation for many years, but he first boldly proclaimed his policy in a seminar at the American University in Beirut in June 1968. It was in the wake of the humiliation of the Six-Day War, when the West was most sceptical of Arab effectiveness and Yamani felt the need for a militant initiative. He pointed out that many of the newcomers to the Middle East had agreed to various forms of partnership, and he warned the sisters: 'the June war, with all its psychological repercussions, has made it absolutely essential for the majors—and not least Aramco—to follow suit if they wish to continue operating peacefully in the area. Partnership with the host governments is a must; and delay will be paid for by the oil companies concerned'.[2] A meeting of OPEC in the same month duly endorsed the principle in a Declaratory Statement, that 'the Government may acquire a reasonable participation'.

But the oil companies were very far from agreeing. Some company men realised that their control could not last, but the sisters could not bring themselves to give anything up before they had to—for the profits were still so vast that every extra day seemed worth hanging on to. Some ammunition to the Arabs was given in 1970 by the U.S. Department of Commerce. It estimated that the net assets of the petroleum industry in the Middle East were $1·5 billion, yielding profits of $1·2 billion, a return on investment of 79 percent—compared to only 13·5 percent from smelting and mining industries in the developing countries. It was a calculation that was carefully noted by OPEC.[3]

Yamani saw the Saudis' position as being now so strong that they could get what they wanted; and in March 1969 he described participation to Richard Johns of the *Financial Times*, as creating a bond which 'would be indissoluble, like a Catholic marriage'. At the same time, he urged that participation would be in the interests of the oil companies. 'It will save them from nationalisation,' he explained at a seminar in Beirut, 'and provide them with an enduring link with the producing countries'. But some of the majors, he observed wryly, 'seem to be obsessed with the empire they have built. It is so vast and it took them so many decades to achieve. And now they see these newcomers—these national oil companies in the producing countries—wanting to come and take a piece of their cake, which is the last thing they want to happen.'[4]

It was not until after the Teheran agreement in 1971, when OPEC was stretching its muscles, that Yamani made much further progress. The Saudis again felt the need to show an initiative, after the success of the Shah, and the two chief oil-producers were more than ever rivals for leadership. The Teheran agreement, binding for five years, appeared to leave no room for argument; but participation provided a field for demands outside the agreement. 'It was like labour unions', said one Dutch Shell director: 'when they've agreed about wages, they go for co-determination.' For the next two years the battle for participation took over from the battle for prices.

At the OPEC conference in June 1971 Yamani succeeded in convincing most other members, against the arguments of Tariki, that they should immediately demand a 20 percent share in the companies' operations, to be gradually increased to 51 percent, with the assets to be paid for at 'net book value'. Later Yamani was put in charge of negotiating with the companies for all the Gulf countries. The companies on their side again resolved to stand together. John McCloy got special permission from the anti-trust chief, McLaren. The seven sisters and others met in London to decide on common tactics, renewing their agreement to share out their oil if one company was singled out.

The two sides were again heading for confrontation: the pace was quickening, and the more radical members of OPEC were determined not to be outdone.

Already in February 1971 Algeria, after bitter disputes about prices, had nationalised 51 per cent of all French interest in her oil, thus dashing French hopes of a special connection. Now ten months later, just as OPEC was meeting again, the Libyans announced the nationalisation of all BP's assets, on the unconvincing grounds that Iran had just occupied three islands in the Persian Gulf that technically were under British protection; and that BP, because it was half-owned by the British government, must take responsibility. The nationalisations encouraged the other producers to be more militant.

In January 1972 OPEC met the companies again in Geneva. OPEC's first aim was to obtain more money because the dollar had been devalued: the companies quickly agreed to link the posted price to a basket of currencies, instead of the dollar, and thus put up the price by 8·5 percent, which cost the consuming countries another $700 million a year for oil from the Persian Gulf alone. Then OPEC went on to demand participation in more detail: the companies soon bitterly complained about the terms, and the new battle began.

The prospect of participation placed the sisters in a much more exposed position, stuck in their no-man's-land between producers and consumers. Their problem was forcefully analysed by the oil consultant Walter Levy in New York, a man who from now onwards was to play an increasingly important role in oil policy. Levy had a very international background: he had been educated in Hamburg and worked as an oil journalist in London before moving to Washington during the war; and then set up as an independent consultant. He was now retained as adviser to all seven sisters, and to several governments. Writing in *Foreign Affairs* in 1971, Levy warned that participation was a grand design, in Yamani's mind, to 'bind the interests of the oil companies.' Levy predicted that 'the oil companies would be destined to become completely subservient to their host government.' They would thus face a major decision: 'to what extent and for how long they can be held hostage by their resource interests in producing countries . . . It should now be recognised that their position as an internationally integrated private industry depends on a closer relationship and better understanding with consuming governments.'

The companies were already, as Sir Eric Drake put it in 1970, the tax-collectors of the producers; now they would be much closer partners. Professor Maurice Adelman of MIT, writing in 1972, accused them of being simply 'agents of a foreign power'.[5] For a few oilmen participation appeared a worse evil than nationalisation, for it would impair their freedom of movement, and commit them to the producers' side: Sir David Barran of Shell (which had no huge Middle East concessions like the Aramco partners) said in April 1972 that nationalisation would be preferable to participation, which he regarded as 'intolerable'. But most of the companies, though they were to argue bitterly over terms, were inclined slowly to give way to participation; for it was still the crude oil that provided most of their profits. And they began to perceive that what mattered in a growing shortage was not so much who owned the oil, as who was able to buy it.

Aramco was now the critical battleground for participation; it was both on Yamani's home ground, and it was by far the richest prize in the Middle East. The four owner-companies— Exxon, Texaco, Socal and Mobil—were more thoroughly dependent on it as the shortage elsewhere was more pronounced, and for each of them Aramco was the source of 'the incremental barrel'. Since the 'fifties the Aramco company town at Dhahran had expanded into a settlement with every indication of permanence. Trees, gardens and two-storey houses had grown up inside the barbed wire, and the 'Aramcons' had become almost a separate breed. There was now a generation of employees who had been born and brought up in the desert. It had become a kind of symbol of the no-man's land of the oil companies, belonging to no culture and no country. Visiting it, I felt as if I were inside an idealised America, like an old cover of the *Saturday Evening Post*; an America untouched by the turmoil of the 'sixties, by long hair or drugs, with its citizens watching old movies on Aramco TV, playing baseball or mending their cars. To the Aramco engineers the compound offered not only fat salaries and early retirement, but a kind of engineer's utopia—progress without politics, and technology without doubts. All around them the twisting pipes, the giant towers and flares were the monuments to their skills.

The Aramcons had done what they could to satisfy the Saudis.

Ever since the early 'fifties, the headquarters of the company had been out in the desert, whither the directors would fly from New York or California. Saudi engineers lived alongside the Americans and hundreds of Saudis were sent to America for training. There were Saudi-pleasing public relations projects, and Aramco experts on Saudi culture, history and politics. The new president, Frank Jungers, had been in Arabia ever since graduating at Washington University, and his son was following him into Aramco. A stocky slow-speaking engineer, Jungers was essentially a pragmatist, talking about Saudi politics in the same matter-of-fact style as he talked about Aramco's great projects — the unique installations, the 'first-time-built size-wise facilities'. His Texan-style house, with romanticised pictures of idealised Arabs on the walls, looked on to a bright green garden, with the desert beyond the wire: Aramco was his life, and Dhahran was his home. Jungers disseminated to the Saudis a quiet confidence in all things American. As one oilman described him: 'He's like W. C. Fields, who dropped out of the sky into the bed of a fair maiden, and reassured her in a state of shock: "Don't worry about a thing ma'am. I'm a citizen of the United States of America." '

Yamani began his negotiations with Jungers early in 1972. The two men got on well, but in the background the owning companies — 'the four neurotic parents' — were standing fast. Yamani was exasperated by their stubbornness: 'It is always hard to be moderate,' he complained to a British visitor in January 1972, 'but it is particularly hard with American oil people, because they are apt to mistake moderation for weakness . . . I am afraid that sometimes, when oilmen think only from a money angle, they get blind.'[6] Aramco put forward a rival proposition, that the Saudis could have a 50 percent share in future developments, which Yamani quickly rejected.

Aramco also rashly tried to recruit the help of Washington, thus breaking the tradition of separation from government. Ken Jamieson, the Chief of Exxon, called on Nixon at the White House, who passed a message to Riyadh. Yamani, furious at the intervention, retaliated by recruiting the support of the King himself. The next day he read a stern message from His Majesty, with a clear threat of unilateral action. 'Gentlemen: the implementation of effective partnership is imperative, and we expect the companies to co-operate with us with a view to reaching a satisfactory settlement.

They should not oblige us to take measures in order to put into effect the implementation of participation.'

Yamani warned that the producers 'must prepare for battle', and convened another extraordinary conference of OPEC. But on the eve, Aramco accepted the principle of immediate government participation of 20 percent. It was now the Americans who were making the mistake that they had attributed to BP in Iran twenty years before; of 'failing to yield gracefully that which they were no longer in a position to withhold'.[7]

The next month Yamani continued his talks on participation in San Francisco with the Aramco sisters and BP and CFP. The diehard companies were still resisting, and Socal bluntly told Yamani that their share in Aramco was not for sale. There were more attempts at U.S. government pressure: John Connally, then Secretary of the Treasury, made a hawkish speech on April 18, saying that the U.S. would soon have to back up private corporations in dealing with foreign governments which controlled natural resources critical to America. The OPEC negotiators promptly warned that any such intervention would greatly complicate the participation negotiations. The State Department agreed with the companies that the Teheran agreement did not allow for participation; and they made representations through the Ambassadors in Iran, Saudi Arabia and Kuwait. But they eventually restricted themselves to protesting about compensation at book value, pointing out that future OPEC investments in the U.S. might meet a similar fate—an issue which was eventually resolved with a complex compromise.[8]

Participation was now the catchphrase of the Middle East. The rich Sheikhdoms of Abu Dhabi, Qatar and Kuwait all soon wrested agreements for 20 percent ownership from the companies. And in Iraq the long arguments with the IPC, the grandfather of all the consortia, were reaching a bitter climax. The Iraqis were demanding a greater share in the revenue, participation in the concessions, and a guarantee of high production; they complained that the companies had been restricting production to punish Iraq for the past decade.[9] Eventually in June 1972 Iraq finally nationalised the consortium, and soon afterwards agreed with the French company, CFP, to take its usual share of the nationalised oil. The Iraqis, in a time of shortage, could now play the French against the seven. The old warning of 'look what happened to Mossadeq' was no longer valid, as the nationalised companies showed they could operate their own oil.

Through 1972 Yamani went on bargaining with Aramco, who still resisted, and the King repeated his warning. Eventually in October the companies met in New York with the five Persian Gulf States, headed by Yamani, and reached a 'General Agreement', that they would give up 25 percent of the established concessions, rising to 51 percent in 1983. The final agreement was signed at the end of the year in Riyadh.

But the Battle for Aramco still raged, now centring on the question of the price the four companies should pay for the oil that they bought back from the Saudi government's share—the so-called 'buy-back price'. Yamani knew that he now held the whip hand, and the four sisters, led by Exxon, could not risk a real showdown. For the Aramco board had now approved a plan to expand vastly the Aramco oilfields to provide for 13.4 million barrels a day by 1976, and as much as 20 million barrels a day by 1983. For this breathtaking prospect, even though it would be shared with the Saudis, the Aramco partners were prepared to yield over prices. Their greater fear was that the Saudis would sell their oil to other companies.

By December 1972 George Piercy of Exxon noted that Yamani was 'prepared to relax and await capitulation' and recommended his colleagues to improve their offers. But Gulf, which was outside Aramco, was exasperated by the other companies' willingness to 'buy back at inflated and irrational prices'. Gulf did not have the same self-contained system of outlets as Exxon or Mobil, and was thus more concerned about price—and about consumers. Their chairman Bob Dorsey protested to the team: 'your insistence on maintaining control of the oil for your own use has given strength to the OPEC position and has brought us to this point'. He went on with asperity to suggest that the prompt acceptance of higher prices 'might not prove acceptable to consumer interests'.[10] But Aramco was desperate to settle their access to the oil, and Gulf eventually had to negotiate an agreement in Kuwait similar to Aramco's.

Yamani could now divide and rule the sisters, as they had once done with him. Mobil, still hungry for crude oil, and more militant under the regime of Warner and Tavoulareas, was determined to get more than its 10 percent share of the 'buyback oil'. They were threatening to make a separate deal with the Saudis if their terms were not good enough, as the Socal negotiator, Jones McQuinn, anxiously observed,[11] and Yamani could use the threat to extract better terms from the others.

The negotiations dragged on through the summer of 1973. Yamani refused to bargain on the buyback price, and at one point in August, he warned the Aramco team: 'don't be surprised if at any moment, I pick up the phone and instruct Brock or Frank[12] to cut production to seven million barrels a day'. At last, on September 13, 1973, the Aramco partners gave in to Yamani's stiff terms—a buyback price of 93 percent of the posted price—which would soon increase the price to the West.

Yamani had got what he wanted—a stake in the great concession. And the companies, having fought so long and ruthlessly, were now more relaxed about the new prospect. For participation now established a common interest between the companies and the producing countries, with both sharing the same interest in the orderly world market. It worked like the fifty-fifty tax deal twenty years earlier, but at a deeper level. The oligopoly of the companies had now been effectively joined by the oligopoly of OPEC.

But to the radical critics the new arrangement was at least as sinister as the old one. Professor John Blair, the compiler of the original FTC report on the Petroleum Cartel, suspected that the industry was now entering a new phase of 'bilateral monopoly', with the producing countries as sellers and the companies as buyers: so long as the companies kept the exclusive right to buy the oil, he believed, they could control the supply, 'which is the heart of the control of the price'. Professor Adelman saw the whole situation after the Teheran agreement as a take-over of the old cartel by the new one: 'The producing countries now have a cartel tolerated by the consuming countries and actively supported by the United States.' 'OPEC,' he warned, 'has come not to expel, but to exploit.'

Certainly as Levy had warned, the companies' role was now much more awkward, much harder to separate from politics. As Jim Akins warned them from the State Department, they might find themselves as minority partners of both producer and consumer governments, with a more circumscribed role in negotiating. 'How the companies react to these pressures,' wrote Akins, 'and what they offer as alternatives, will to a large extent determine their future form and their future activities'. In fact, they did not

react. Pressed on both sides, they hoped for the best, and for the
time being they seemed to co-exist peacefully with the producers.

But even as the Aramco agreement was being prepared a new
crisis was brewing up in the Middle East; and before it was signed,
Yamani had left the U.S. in a hurry. A chain of events had been
started which was to mix up the question of prices inextricably
with the question of politics, a mixture made more dangerous
by the growing world shortage of oil.

SHORTAGE

Up till the late 'sixties, the sisters were worried about having
too much oil, not too little. In May 1967 Michael Haider, then
chairman of Exxon, answered a shareholders' question at the
annual meeting in Houston with the words: 'I wish I could say
I will be around when there is a shortage of crude oil outside the
United States.' A year later Socal in California was specially
worried about the Alaska discoveries, in which they had no share.
A Socal company memo in December 1968 warned that 'within
five to ten years there may be large new crude supplies from the
Arctic regions of the world seeking markets and thereby extending
and magnifying the surplus supply problems.'[13]

In 1969 George Piercy of Exxon was fairly confident that his
company's future supplies would meet demands. On the one side
he expected that the boom in Japan—the biggest single importer
of oil—would begin to slow down. On the other hand there were
the huge new expected sources of oil, including Alaska, Libya
and eventually the North Sea. But already there was an ominous
turn-down in Exxon's master-graph, showing spare capacity compared to world demand. By 1970 the trend was much more serious.

The world's demand for oil was ahead of all the predictions;
a memo from Gulf in March 1970 pointed out that their estimates
of two years earlier were 8 percent lower than the actual consumption: 'if once again our estimates of future free world demand
prove low, then a strain on productive capacity may be approached
before 1980'. Was this consistent underestimation always a genuine
statistical error, or was it sometimes inspired by the companies'
instinctive fear of a glut?

Within the United States, production of oil was no longer going
up, and after 1970 it went down. For the first time the optimism
of the early drillers that there would always be oil somewhere else

was now unfounded. Already by 1970, 28 percent of the oil used in the U.S. was imported. The possible danger of this to national security prompted President Nixon to appoint a special Cabinet task force, headed by George Schultz, which reported in 1970 with historic complacency, against the advice of the oil companies,[14] that there was little danger of an Arab boycott, and that existing import controls should be liberalised. The president did not accept the report, but import quotas were nevertheless relaxed, and the imports of Middle East oil went up and up.

The oil-producers had already begun to realise that their bargaining position was stronger, as Yamani had warned Piercy in February 1971 ('George, you know you cannot take a shutdown'). By 1972 many experts reckoned that the world was heading for an acute oil shortage in a few years, and Shell was sending out serious warnings. In October 1971, Barran of Shell warned that the days of cheap oil were over, and that by the end of the century oil consumers would be 'looking down the muzzle of a gun.' There was a new danger sign when Kuwait decided in 1972 to conserve its resources and to keep its production below 3 million barrels a day.

For BP and Gulf, the two partners in Kuwait, this was menacing news. But Exxon, like the other American partners in Aramco, was not seriously worried, for they were confident that Saudi Arabia could supply all the extra required. By early 1973 the Saudis were producing $6\frac{1}{2}$ million barrels a day, and the prospects were even more dazzling. As other parts of the world became more uncertain, so Saudi Arabia became more crucial.

After the companies had agreed to Yamani's participation terms, it soon emerged that they would push up prices. For there was now great demand for the relatively small amount of oil that the producing countries were selling on the free market. And by the beginning of 1973 the price in the free market was rapidly rising.[15]

In April 1973 a new warning appeared in the august pages of *Foreign Affairs*, much more alarming than Levy's two years before, from none other than Jim Akins, now increasingly outspoken. He presented the ominous statistic that world consumption of oil for the next twelve years was expected to be greater than the total world consumption of oil throughout history up till 1973. The loss of production from any two Middle East countries could cause a panic among consumers. The price of oil was likely to go up to $5 a barrel well before 1980. The Arabs could use oil as a political weapon, for the advanced countries were obviously now vulnerable to boycott.

Akins' warnings were prophetic, but some of his critics insisted that they were self-fulfilling. The State Department was virtually advising the Arabs to put up prices, and advertising the West's weakness. Adelman, writing in the fall of 1972, had insisted that the talk of shortage merely reflected the interests of the big oil companies, in cahoots with the Arabs. There was 'absolutely no basis to fear an acute oil scarcity over the next fifteen years'.[16] Adelman, like many other oil experts at the time, was confident that an Arab embargo could not be sustained, as the fiasco of 1967 had suggested.

But the Arabs hardly needed Akins' advice. For the signs of a shortage were now visible everywhere. The summer of 1973 was an eerie one for the oil companies. The demand for oil was going up above the wildest predictions — in Europe, in Japan, and most of all in the United States. Imports from the Middle East to the U.S. were still racing up: production inside the United States was still falling. In April President Nixon had again lifted restrictions on imports of oil, so that Middle East oil flowed in still faster; and the administration did nothing to control a scramble for oil. While the majors were trying to establish their safe sources of supply, the independents were bidding frantically for the 'participation oil' from the producers, thus pushing the price up and up.[17]

In the gathering crisis, some oil companies tried to involve the consumer governments. Sir Eric Drake in the BP annual report of 1972 said there was an urgent need for consuming governments 'to adopt coherent, balanced and co-ordinated energy policies'. And in June 1973 Frank McFadzean of Shell told a seminar at Harvard that the energy problem was now a question of political power which required government intervention.[18] The State Department, urged by Jim Akins, had tried periodically since 1971 to arrange joint talks with the Europeans, without success, but it was not till June 1973 that OECD set up a group to discuss emergency oil policy, to report in November: and it was too late. The prospects of serious coordination across the Atlantic and Pacific were anyway slender: each continent had a different dependence on oil, and a different attitude to the Arabs and Israel. The more political the oil, the more it divided the West.

Yamani and the other oil ministers soon became aware of their new opportunity. In June 1973, as prices were zooming up, OPEC summoned another meeting in Geneva, to insist on a further increase because of the further devaluation of the dollar. The militants — Algeria, Libya and Iraq — were now pressing for unilateral control of price, but eventually OPEC agreed on a new formula which put up prices by another 12 percent. One evening after midnight Yamani went on a jog through Geneva with three correspondents, and predicted that this would be the last time prices would be negotiated with the companies.

In early August Yamani warned Aramco that the Teheran agreement would have to be renegotiated well before its time limit of 1975. The next change would be a very large one and there would be no real negotiation: 'there will be discussion within OPEC,' Yamani told them, 'with the wild ones insisting on a very high level. And there will be a compromise within OPEC, but then the companies will have no choice.'[19]

By September 1973, for the first time since OPEC's beginning, market price of oil had risen above the posted price. It was a sure sign that OPEC were now in a very strong bargaining position. The glut that had weakened and divided them since 1960 was now emphatically over. Armed with this knowledge, OPEC invited the companies to meet them in Vienna on October 8 to discuss 'substantial increases' in the price of oil.

THE KING'S MESSAGE

While the shortage loomed, the Arabs were at last achieving closer unity. They were determined to use oil as a weapon against Israel, and by 1973 the militants were being joined by the country on which the four American sisters had pinned all their expectations for increase. The very fact that Saudi Arabia was now far the biggest oil exporter made King Feisal more vulnerable in the face of his Arab colleagues, and the danger of an embargo more likely; for he could not afford to be seen as a blackleg.

The oil companies became well aware of Feisal's worsening predicament but their attempts to warn Washington were met with scepticism. As with the shortage, it was a case of Wolf, Wolf. The oil lobby, which had always been so ready to invoke the national interest to protect profits, were now *really* dealing with the national interest — but it was disbelieved or ignored. It was an ironic consequence of the State Department's policy, proclaimed twenty years before, that 'American oil operations should be the instruments of foreign policy in the Middle East'. They had now delegated that policy so completely that when the warnings came, the bells did not ring.

Ever since the Six-Day War, the Aramcons in their compound in Saudi Arabia had become alarmed by King Feisal's growing concern over Israel. Frank Jungers and his colleagues carefully briefed all visitors to the camp, ranging from Senator Gravel to General Goodpaster, about the depth of Arab feelings and the dangers of the U.S. foreign policy. 'The story of the senseless dissipation of the goodwill we used to have among the Arabs.' said one of many such briefings, 'constitutes one of the saddest chapters in the history of our foreign relations.'[20]

At the beginning of 1973 the King was taking some trouble to influence Washington, both through Aramco and through Akins, who provided the chief link between the Arabs and Washington. But Akins had to go to extreme lengths to convey the Saudis' views: in January 1973 John Ehrlichman, President Nixon's aide at the time, was preparing a visit to Saudi Arabia. Akins asked Aramco to arrange for Sheikh Yamani to 'take Ehrlichman under his wing and see to it that Ehrlichman was given the message: we Saudis love you people but your American policy is hurting us.'

On May 3, 1973 Jungers paid a courtesy call on King Feisal, for half an hour. The King was cordial but his tone was quite different to that of earlier meetings. The King touched only briefly on his usual hobby-horse of the Zionist-Communist conspiracy, but he warned Jungers that Zionists and Communists 'were on the verge of having American interests thrown out of the area'. Only in Saudi Arabia, the King stressed, were American interests relatively safe; but even in his kingdom it 'would be more and more difficult to hold off the tide of opinion'. The King was amazed that Washington failed to perceive its own interests: 'it was almost inconceivable in any democratic state' (he told Jungers) 'for a government to be so far away from the interests of its people'. But it was easily put right, he went on: 'a simple disavowal of Israeli policies and actions by the U.S. Government would go a long way...'[21]

Jungers then went on to see Kamal Adham, the King's chamberlain and close adviser, who gave a more ominous message. The Saudis he said, in spite of their problems with the Egyptians, could not stand alone when hostilities broke out. Adham was sure that Sadat, a courageous and far-sighted man, would have to 'embark on some sort of hostilities' in order to marshal American opinion to press for a Middle East settlement. It was a very specific and accurate warning on top of the King's audience: 'I knew he meant war', said Jungers later. 'The King liked to give signals, first subtly and then explicitly. It was quite different from his earlier warnings.'[22] Jungers quickly passed on the warning to Exxon and Co. in New York and California.

Three weeks later the four Middle East directors of the parent companies—Hedlund of Exxon, Moses of Mobil, Decrane of Texaco and McQuinn of Socal—were at the Geneva International Hotel for a meeting with Yamani to negotiate about participation. Yamani suggested that they might pay a courtesy call on the King, who had just been to Paris and Cairo, where Sadat had given him 'a bad time' (as Yamani put it) pressing him to step up his political support. The King, after a few pleasantries, was much more curt and abrupt than usual. He warned the four directors that time was running out. He would not allow his kingdom to become isolated, because of America's failure to support him, and he used the phrase 'you will lose everything'—which to the visitors could only mean that their oil concession was at risk. The King asked them to make sure that the American public were told where their true interests lay, instead of being 'misled by controlled news media'.

The Aramco men lost no time, and a week later they were all four of them in Washington to lobby top government officials. To each they repeated the King's message, that unless action was taken urgently, 'everything would be lost'. On May 30 they called first on the State Department, to see a team led by Joseph Sisco, in charge of Middle Eastern affairs. But Sisco had heard such warnings before, and he assured them that his information was otherwise. The CIA had reported, through its own contacts including close relatives of the King, that Feisal was only bluffing: he had resisted pressure from Nasser in the past, and could resist pressure from Sadat now.

They then went to the White House, hoping to see Kissinger; but they were fobbed off with General Scowcroft, and other advisers including Charles Debono, then the energy expert. Finally they went to the Pentagon to see Bill Clements, who was then acting Secretary of Defence, while James Schlesinger was awaiting confirmation. Clements had been an oil man himself, with his own drilling company, and was widely regarded in Washington as a key figure in the oil lobby. But he made clear to his visitors that he had his own information and views about the Arabs: they would never unite, the companies' fears were unfounded, and King Feisal was dependent on America. At the end of their day in Washington, the Aramco men cabled sadly back to Jungers in Saudi Arabia that there was 'a large degree of disbelief' that any drastic action was imminent. 'Some believe that His Majesty is calling wolf where no wolf exists except in his imagination.'[23]

But the four companies still wanted to show the King that they were trying to influence American opinion. Each wanted to protect their future share of the concession, and they soon vied with each other to show their helpfulness. Bill Tavoulareas, the president of Mobil, was a close friend of Yamani, and he personally lobbied Sisco at the State Department; but Sisco again was sceptical. Mobil also prepared an advertisement for the *New York Times* on June 21. It was very cautiously worded: it explained how America was becoming increasingly dependent on imports from Saudi Arabia, how relations were deteriorating and how 'political considerations may become the critical element in Saudi Arabia's decisions'. It concluded that it was 'time now for the world to insist on a settlement in the Middle East.' But the *New York Times* thought it too inflammatory for the usual position for Mobil's advertisements, opposite the editorial page. In Saudi Arabia, nevertheless, the advertisement had the required effect: Yamani wrote a letter to Mobil recognising this 'positive step'.

Exxon was rather more discreet in their support. They decided against advertising, but Ken Jamieson pressed their case in Washington, and Howard Page gave a speech in New York to alumni of the American University in Beirut, about the need to relieve the strained political relationships between the United States and the Arab countries. In California, Socal now became worried that *they* were slipping behind, and their Foreign Review Committee was concerned that 'Socal will be conspicuous in our absence'. Consequently Socal's very conservative chairman, Otto Miller, wrote a letter to shareholders on July 26, urging that the U.S. should work more closely with the Arab governments, and 'acknowledge the legitimate interests of all the peoples of the Middle East . . .' The letter was well-publicised, and caused a small furore among Jewish communities, especially in San Francisco.

There was also plenty of activity from old Jack McCloy, now

seventy-nine, and still representing all seven of the sisters. He talked to his friends in Washington: he warned Sisco that the Saudis meant what they said, and he urged Kissinger to try to mediate. 'I kept jumping on him,' he told me, 'to say that it was an imperative of statesmanship to get the Middle East settled; that the administration mustn't just think in terms of the next New York election'.

The companies were certainly persistent enough. Why, then, did they have no discernible effect? Had not the oil companies given at least $2.7 million to President Nixon's campaign? Had not Gulf Oil been contributing millions of dollars, including a secret gift of $100,000 in 1971, with the express understanding that they would be 'on the inside track'? Yet, when four huge global companies, with billions of assets behind them, wanted to pass a critical message from King Feisal they apparently had no influence whatever.

There were several explanations. The Israeli lobby was undoubtedly far stronger, and American intelligence about the embargo and the war was heavily influenced by the Israelis. Secondly the administration, having for so long separated the two strands of Middle East foreign policy, still kept them in different compartments. But thirdly, the American oilmen—as some of them wryly admit—had lost nearly all their credibility. When the companies did have something serious to say, hardly anyone believed them.

In the meantime the oil weapon was gathering support among the other Arab states. In May, Dr. Nadim Pachachi, the ingenious Iraqi in exile who had been secretary of OPEC, put forward a new political proposal which took advantage of the shortage: he suggested that the supply of crude oil to the West should be frozen to enforce Israel's withdrawal from the cease-fire lines of 1967. The next month the Libyans set a new pace of militancy. Colonel Qadaffi nationalised Bunker Hunt's concession, saying that the United States deserved 'a good hard slap on its cool and insolent face'.

President Sadat of Egypt, instead of moving closer to the Libyans, now significantly altered his alignment. At the end of August 1973 he flew to Saudi Arabia for a secret visit to King Feisal. The meeting was momentous. Feisal promised Sadat that, if American policy in the Middle East did not change, he would restrict the increases of oil production to 10 percent a year—far short of Aramco's requirements. Thus Egypt, for the first time, had oil pressure behind her diplomacy.

Just after the visit, Qadaffi, celebrating the fourth anniversary of the Libyan revolution, announced that he would nationalise 51 percent of all the oil companies operating in Libya, including the subsidiaries of Exxon, Mobil, Texaco, Socal and Shell. Two days later the Libyans announced that the price of Libyan oil would go up to $6 a barrel—nearly twice the Persian Gulf price—and threatened to cut off all exports to America if Washington continued to support Israel. Soon afterwards all ten foreign ministers of the Arab oil exporting countries (OAPEC) met to discuss the possible use of oil as a weapon to change American policy. A fortnight later Sheikh Yamani formally warned the United States that there could be a cutback of Saudi Arabian oil.

President Nixon appeared on television to warn the Libyans of the dangers of a boycott of oil, reminding them of the experience of Mossadeq in Iran twenty years before. But oil experts knew that the threat was hollow. As Ian Seymour asked in the *New York Times*: 'could it really be that the President of the U.S. had not yet grasped the predominant fact of life in the energy picture over the coming decade, that the problem is not whether oil will find markets, but whether markets will find oil?'[24]

NOTES

1. Leonard Mosley: *Power Play*, London, 1974, p. 395.
2. *Middle East Economic Survey*, June, 1968.
3. Abdul Amir Kubbah: *OPEC Past and Present*, Vienna, 1974, p. 78.
4. 3rd Seminar of Economics of the Petroleum Industry, American University of Beirut, 1969.
5. *Foreign Policy*, New York, Fall 1972.
6. Mosley, pp. 400–402.
7. Dean Acheson, quoting Burke: see Chapter 6.
8. James Akins: *Foreign Affairs*, April 1971.
9. *The Nationalisation of IPC's Operations in Iraq*, Baghdad, 1974, p. 8.
10. Multinational Report: 1975, p. 136–7.
11. Ibid: p. 139–40.
12. Brock Powers, chairman of the board of Aramco, and Frank Jungers, president of Aramco, see Multinational Report, p. 138.
13. Multinational Hearings: 1974, Part 7, p. 360.

14. See submissions by Exxon, Texaco, Mobil and Socal: Investigations Subcommittee, Jan. 21, 1974, p. 184–194.

15. The 'rise in the market price,' commented *Petroleum Press Services* in November 1973, 'is partly accounted for by the disruption of normal trading relations following the participation agreements. It was the high prices obtainable by state companies for participation crude—equivalent to only about $2\frac{1}{2}$ percent of the Middle East market—that whetted the appetites of host governments.'

16. *Foreign Policy*, New York, Fall 1972.

17. See testimony of Dillard Spriggs, Multinational Hearings 1974, Part 4, p. 61.

18. *Petroleum Intelligence Weekly*: April 23, 1973.

19. Company cable to New York: Multinational Report, p. 148.

20. Multinational Hearings: 1974, Part 7, p. 524.

21. Multinational Hearings: Part 7, p. 506ff.

22. Interview with author, February 1975.

23. Multinational Hearings: Part 7, p. 509.

24. *New York Times*, October 7, 1973.

12

Embargo

With the possible exception of Croesus, the world will never have seen anything like the wealth which is flowing and will continue to flow into the Persian Gulf.

James Akins, April 1973

In this atmosphere of crisis, the stolid oil company delegates prepared to confront OPEC in Vienna on October 8. The companies were represented by a team of five, led by George Piercy, the Exxon engineer, and André Bénard, the French director of Shell, a former resistance hero who had only recently become involved with the Middle East. They had all been briefed beforehand by their own boards, and by a special meeting of the London Policy Group—once again with anti-trust clearance. Governments were kept informed, but in spite of the growing tension, no diplomats went to Vienna. The negotiation was still between countries and companies. The buffer was still assumed to be working.

And then, just as they were leaving for Vienna, Egypt and Syria invaded Israeli-occupied territory. There was war. It was now all too clear why King Feisal's warnings had been so emphatic: his lonely loyalty to the United States was now untenable.

OPEC, represented only by the Persian Gulf countries, six of the thirteen,[1] was led by the oil ministers of the two biggest producers: Yamani from Saudi Arabia and Amouzegar from Iran. They were in a strong position. The 'house of cards' of the Teheran agreement had now clearly collapsed. Firstly, inflation had galloped ahead, so that the increase of 2½ percent a year was now obviously inadequate: the price of other commodities which the OPEC countries had to import (as the Shah never ceased to point out) was far outstripping the oil price. Secondly the shortage had transformed OPEC's bargaining position: some independents were already offering $5 a barrel for 'participation' oil. Thirdly, the whole psychological balance of power was changing in the face of the shortage. As Dr. Amouzegar put it: 'we became so disgusted by their mercenary approach, bickering over a penny. But we realised that they were in a different position after two wars and the closing of the canal. All these things made them get a little panicky about not getting oil for their customers; we began to feel they were losing their strength.'[2] On top of these factors, there was now the war. The company delegates arrived to find the Arabs in a state of excitement, passing round newspaper articles and photographs of American supplies to Israel.

The negotiations opened ominously. Yamani began by demanding increases in the government 'take' which would, in effect, double the posted price to $6 a barrel. Piercy replied by offering an extra 15 percent. The oilmen talked strictly commercially, in terms of the market as they saw it: 'We argued very effectively,' as one team put it to me, 'perhaps too effectively. We left them with no resort to logic—only to power.' But the OPEC delegates were convinced that, with the shortage, the market was on their side. Yamani made a few concessions, taking the posted price to around $5 a barrel; but Piercy, in terms of his brief from his masters, could not go beyond 25 percent.

In the meantime the war was raging, the Egyptians had crossed the Suez Canal, and the Arab states had been forged into a new unity. The Arab members of OPEC, through their own organisation OAPEC, announced that they would meet in Kuwait to discuss the use of oil as a weapon of war.

The question of the price was now in imminent danger of being confused with the question of an embargo. It was feared not only by the company delegates, but by Yamani, at least by his account: in Geneva four months earlier he had said he would not negotiate again, but now apparently he wanted to avoid a breakdown. 'I wanted to keep the companies with us, while we went into this political period,' he told me later, 'I didn't want to mix prices with politics'. Amouzegar claimed afterwards that he had genuinely wanted to settle: 'We asked for a dollar a barrel of government take, knowing that they would be mercenary as usual, expecting to settle for forty or forty-five cents. They made a terrible mistake. They flatly refused, saying the oil market didn't justify it. But we knew that the low price was due to the world being divided between rich and poor.'

The company delegates cabled to their principals in London and New York and received their replies: they could not agree to such huge increases without consulting the consumer governments. Some of the company delegates regretted their instructions, believing that it was still possible to negotiate an agreement which might otherwise be forced on them. After midnight, Piercy and Bénard paid their momentous call on Yamani at the Intercontinental Hotel, to tell him that they must adjourn for two weeks to consult governments.[3] Yamani made clear his regrets. The delegates could not budge. The next morning some of the other OPEC delegates arrived at the offices to find that Yamani and the company delegates had already flown home. The news of this historic breakdown, which was to transform the economy of the West, was buried in the war news. It was delegated to the business pages, and even in the *Financial Times* it could only reach page 13.

In New York the Aramco delegates were now desperate about their companies' survival, and they turned once again to Jack McCloy to press their case. McCloy promised to get a letter to the President, and wrote to General Haig at the White House on October 12 by special messenger, enclosing a memorandum to be passed on to both Nixon and Kissinger. The memo was signed by all four chairmen of the Aramco group—Jamieson, Granville, Miller, and Warner. They emphasised that any increased American military aid to Israel 'will have a critical and adverse effect on our relations with the moderate Arab producing countries'. The Europeans could not face a serious shut-in, and 'may be forced to expand their Middle East supply positions at our expense'. Japanese, European, and perhaps Russian interest might well supplant United States presence in the area, 'to the detriment of both our economy and our security'.[4] It was a characteristic mixture of military and commercial arguments: American oil must be protected from both European competition, and a Russian invasion.

General Haig took his time in dealing with the letter; three days after its delivery, he replied to McCloy ('Dear John') saying 'I will see that the letter is forwarded to the President and to Secretary Kissinger'. But in the intervening weekend Kissinger and Nixon, alarmed by reports of Soviet shipments to the Arabs and the Israeli predicament, had authorised the immediate airlift of military supplies to Israel.

The next week, on October 16, four Arab foreign ministers arrived in Washington, led by the most crucial, Omar Saqqaf, then Foreign Minister of Saudi Arabia. Saqqaf was in a very bad mood: Nixon had not been able to see him that day, and at a press conference one American reporter said that the Saudis could drink their oil: Saqqaf replied bitterly, 'All right, we will'. The next day, after a last-minute warning from Aramco, Nixon and Kissinger agreed to see Saqqaf and the other three foreign ministers. Saqqaf gave Nixon a letter from King Feisal, stating that if the United States did not stop supplying Israel within two days, there would be an embargo. But Nixon explained that he was committed to supporting Israel; and on the same day the senate voted, two to one, to send reinforcements.

In Kuwait in the meantime the members of OPEC had assembled for a double-barrelled attack on the West. On the first day they decided unilaterally to raise the oil price to $5.12 a barrel—70 percent more than the figure of $3.00 agreed at Teheran, and an increase that made all others look tiny. On the second day the Arabs met by themselves, for the meeting of OAPEC. They agreed on an immediate cutback in oil production of 5 per cent, and warned in their communiqué (which was issued only in Arabic) 'that the same percentage will be applied in each month compared with the previous one, until the Israeli withdrawal is completed from the whole Arab territories occupied in June 1967 and the legal rights of the Palestinian people restored'.

The two cuts together—the price-hike and the embargo— proved a deadly combination to the West. But the coincidence, surprisingly enough, was accidental. The embargo had been conceived solely in the context of the Arab–Israel conflict: 'It had nothing to do with wanting to increase the price of oil,' the Secretary of OPEC, Ali Atiga, insisted later: or with increasing the power of the oil producers. It was meant simply to attract the notice of the public in the West to the Israeli question; to get them to ask questions about why we did it.'[5] But the embargo turned out to be the most effective possible device to jack up the price still further. It made not only the Arab producers, but the Iranians and Venezuelans who had no hand in it, aware of their true position of strength.

Three days later, after Saqqaf had returned angrily from America, came the bombshell. Saudi Arabia, far from distancing herself from OPEC, announced a cut-back of 10 percent, plus an embargo of all oil to the United States and the Netherlands. The OPEC ring had now closed tightly and the country which Americans had regarded as the safest was now the most extreme. The next day, the Saudis cut back production by more than 20 percent, and in Riyadh Yamani summoned Jungers to discuss the implementation of the embargo.

The Saudis had already worked out the embargo in some detail. They insisted that on top of the 10 percent cutback, Aramco must subtract all shipments to the U.S. including the military. Aramco would have to police the whole complex operation, and any deviations from the ground rules would be harshly dealt with. Jungers pointed out some of the effects of the rules; for instance that Italy and Japan would be specially hard hit. Yamani remarked that this was deliberate, implying that they were being punished for pro-Israeli attitudes. If this embargo didn't change American policy, Yamani explained, the next step would 'not just be more of the same': Jungers had no doubt that he meant complete nationalisation, if not a break in diplomatic relations.[6]

The Aramcons now found a real crisis of identity; for where was their ultimate loyalty? At the height of a war, with American public opinion highly-charged, they were required to be the agents of an Arab government in enforcing an embargo, including a military embargo, designed openly to change United States foreign policy. The whole past justification for Washington's diplomatic support of the companies in the Middle East—that they were essential to national security—was undermined, just when the U.S. Sixth Fleet was being held in readiness.

By October 20, Saudi Arabia had embargoed all oil shipments to the United States. The next day in New York the Security Council unanimously adopted the call for a cease-fire by the two superpowers. The Aramco directors heaved sighs of relief. 'No question that Saudi Arabian Government mood now more relaxed', Jungers cabled to Socal on October 24, 'but one of cautious anticipation.' But, he warned, the King was still radical on the point of Jerusalem—'always the most sensitive and uncompromising issue with His Majesty'—and there was 'absolutely no question that the oil cutback would remain in effect until the entire implementation was worked out.' One Saudi contact told Jungers that there was 'great satisfaction' with Aramco for its pro-Arab stand, and remarked 'we hope to reward you'—which Jungers interpreted as a promise to allow future growth.[7] The vast project for Aramco's expansion was now a multi-billion political hostage.

Aramco was now closely in touch with Jim Akins, who had

just become the Ambassador to Saudi Arabia. On October 25 Akins sent a confidential message to Aramco, to ask the oil tycoons in America to hammer home to their friends in government that oil restrictions would not be lifted 'unless the political struggle is settled in a manner satisfactory to the Arabs'. There were 'some communications problems', Akins pointed out, with considerable understatement, between the industry and the government. The oil companies must put their views in an unequivocal way. Akins' message was duly transmitted the next day to the four Aramco partners in New York and San Francisco. It was an odd reversal of the diplomatic process. Was it the companies who were the instruments of the State Department, or vice-versa?

The Aramco directors still had to carry out the King's orders, even if they were directly against their own country's interests. The Saudis asked for details, within two days, of all their crude oil used to supply the American military bases throughout the world. In New York officials from Mobil and Texaco were reluctant to release it. But Exxon consulted with the Defence Department (though in a very peremptory fashion: Admiral Oller, in charge of Defence Fuel, did not hear about it until it was a *fait accompli*).[8] The details were provided, and the Saudis duly instructed Aramco to stop the supplies to the military. The position was serious enough for a managing director of BP in London to receive a phone call from Washington asking whether BP could supply the Sixth Fleet. Exxon and the others were now wide open to the charge that had so often been made against them in the past—that they put profits before patriotism.

In New York, it was all too evident that Aramco was carrying out the instructions of a foreign government, and however much they insisted that they had no alternative, they had few public figures on their side. In the midst of the most Jewish city, they stood out as a pro-Arab enclave. The topography seemed symbolic: on one side of Sixth Avenue in Manhattan stood the three skyscrapers of the three TV networks—CBS, NBC, ABC—all of them sympathetic to the Israelis, and deeply critical of the oil companies. On the other side stood the headquarters of the two key companies: Exxon's new skyscraper and the two floors of Aramco, at Fifty-Fourth Street. It was as if the Avenue was an impassable frontier, like the River Jordan itself. Inside the Aramco offices there were young Arab trainees, pictures of King

Feisal, engravings of Arabia, and rows of those bleak colour photographs of pipelines in the desert, so much favoured by oilmen.

To the media, across the avenue, Aramco appeared an all-powerful supra-government, a consortium of four of the richest companies in the world in league with an alien sovereign state. But the Aramco men in New York saw themselves as persecuted and encircled; anonymous telephone callers rang up with bomb scares to make them troop out of the building, and with threats and insults against the oil traitors. There was little middle ground: each side had its own view of the priorities of foreign policy, and each had a profound distrust of the other. The opposite interests had now come together in an explosion.

THE PRICE

The embargo, having originated quite separately, was now rapidly exacerbating the shortage, and thus affecting the price. The oil producers, through their participation, now had oil to sell for themselves on the market, which enabled them very soon to get still higher prices. The oil companies had often in the past exploited a political crisis to push up their prices; the American companies had done so seventeen years earlier, after Suez. But OPEC, being now a more effective cartel, was better able to take effective advantage.

The companies, surveying the situation with alarm and confusion, asked to meet again with OPEC, and on November 17 they assembled in Vienna—this time not to negotiate, but to discuss. There followed a bewildered discussion about the state of the market, for neither side altogether understood what was happening. Yamani insisted that the market was the only thing that he believed in: why should the oil companies say that the market worked when prices were depressed over the previous decade, and now reject the market at a time of shortage? Was it not a paradox that the oil companies should now ask for a substitute for the market price? The companies insisted that the embargo was not letting the free market work. Dr. Amouzegar of Iran, however, was already insisting that the posted price was too low, pointing to the huge excise taxes that the consuming countries continued to levy on oil.

OPEC then held its own conference in Vienna, to discuss how to determine on oil prices. They decided that their Economic Commission should report on the question in December. The companies, their communiqué said, had failed to make a constructive proposal, and the statement continued in OPEC's most reproving style: 'The companies' representatives dwelt vaguely on ideas for pricing of petroleum on the basis of a rigid and arbitrarily pre-determined procedure divorced from normal market forces. The conference is not in agreement with such an approach and believes that the pricing of petroleum, like the pricing of other internationally-traded manufacturered goods, commodities and raw materials, should be market-orientated.' It was a comic reversal of roles: OPEC, having discovered its monopoly power, had become the champion of the market-place: the producers had not only stolen the companies' cartel, they had stolen their humbug.

In the panic and shortage caused by the embargo, the majors were now terrified that the independents—who were much the most desperate for oil, not having their secure supplies—would bid up the price to ridiculous levels. In London, the companies tried to get Peter Walker, the Minister for Trade and Industry, to refuse foreign currency to bids above $6 a barrel, but he would not agree. The companies were specially worried by Japan, who was most dependent on oil: Frank McFadzean of Shell asked the Japanese prime minister, Kakeui Tanaka, to try to restrain bids, but without success.

In Iran on December 16, the Iranian State Oil Company, NIOC, for the first time conducted an auction of some of its oil, with horrific results for the West. The highest bid came to seventeen dollars a barrel. Most of the high bidders were independents, but one bid of twelve dollars (according to Amouzegar) was put in by a Shell affiliate. The majors could not stop the bidding, and the results supported the Iranian argument, that the posted price was far too low.

The climax came just before Christmas. On December 22 the six Persian Gulf members of OPEC met again, this time in Teheran, an ominous setting. For this was the Shah's home ground and the Shah had come to realise that the embargo, though he had played no part in it, was giving him the chance of a lifetime to overcome all his economic problems. Ever since the Teheran agreement of 1971, he had realised that the power of the oil kings was waning; but (as he candidly admitted to me) it was not till the embargo that he understood the real weakness of the companies.

The Shah had already prepared the ground. In October he had sent out a team to investigate the cost of alternative fuels, to try to establish a fair price for oil: they looked particularly at the United States, West Germany and South Africa (with its extensive oil-from-coal operations). The experts' report gave the Iranians more powerful arguments for conservation or, in other words, higher prices. As Dr. Amouzegar recalled it, 'we were specially struck by the fact that in 1951 coal accounted for 51 percent of fuel in the United States, while in 1973 it was 19 percent. Because of cheap oil, all alternative sources were being neglected. We realised that no-one in the West was worrying about what would happen when the oil ran out, and the communists could easily take advantage'.[9] The Iranians could thus smoothly argue that expensive oil was really in the best possible interests of the West. And the Iranian arguments were supported by the findings of OPEC's economic commission, whose report just before the conference indicated that the price should be around seventeen dollars. The prices at auctions encouraged the claim; on the very day of the OPEC meeting an auction in Nigeria had produced a new record bid of $22.60 a barrel—though it was never actually paid.

When the OPEC ministers met, the Iranians and the Saudi Arabians were at loggerheads. The Shah wanted a price of fourteen dollars a barrel which was, he insisted, less than the demands of some others. Sheikh Yamani, on the other hand, was seriously worried by the effects of another big price increase on the world economy. How seriously he was (or is) worried is much disputed by oilmen; many argue that it was his insistence on participation and a high buy-back price which began the whole price escalation. But both Yamani and his master, King Feisal, showed signs at the time of being apprehensive: and Yamani said afterwards: 'I was afraid the effects would be even more harmful than they were, that they would create a major depression in the West. I knew that if you went down, we would go down.'[10]

Yamani knew that the Shah would demand a high price, with the backing of the OPEC militants. The Saudis, as he saw it, were faced with a simple choice: either stick out for a lower price and effectively break up OPEC; or else go along with the majority, and try by all possible means to bring the price down, then and later. 'It was,' he recalled afterwards, 'one of the critical moments of my life; one of the few decisions I took reluctantly.' He could not get in touch with King Feisal, and had to take the decision alone. He encountered two friends in the lobby of the Intercontinental Hotel, and sought their advice. One of them, Bill Tavoulareas of Mobil, advised him to keep the price down, and if necessary break OPEC. The other, an independent oil expert, advised him to fight as hard as he could for a low price within OPEC, but not to break with the majority—a decision to which Yamani was already tending. But it was not, as he discovered later, the King's view and on his return to Riyadh Feisal reprimanded Yamani for not having insisted on a lower price. What would have happened if Yamani had got through to the King that night in Teheran? OPEC would probably have been broken, and the price would have stayed down; but the political isolation of the Saudis would have been increasingly perilous, and it was unlikely that they would have been able to hold out alone.

The next morning the OPEC ministers were still formally meeting when the Shah gave a press conference, to leave no doubt that it was he who was in charge. He announced the shattering increase to $11.65 a barrel, more than doubling the posted price of oil. His assurance was breathtaking: the price, he explained, was very low compared to the going market price, and was reached 'on the basis of generosity and kindness'. The companies, he explained, were making excessive profits of at least a dollar a barrel, for no reason at all. As for the Western consumers, it would do them good to have to economise: 'eventually all those children of well-to-do families who have plenty to eat at every meal, who have their own cars, and who act almost as terrorists and throw bombs here and there, will have to rethink all these privileges of the advanced industrial world. And they will have to work harder . . .'[11]

The OPEC communiqué calmly concluded by saying that, considering that the increase was moderate, 'the Ministers hope that the consuming countries will refrain from further increase in their export prices'.

The price of oil had now quadrupled in just over two months. With incredible suddenness, the whole oil cartel had apparently fallen into the hands, not of seven companies, but of eleven countries. All the converging trends of 1973—the movement towards participation, the shortage, the Arab-Israeli war—had come together to clinch the cartel. The Middle East leaders, in private, expressed their gratitude to the sisters for making it possible. 'We just took a leaf out of our masters' book,' as one Kuwaiti put it. Or as the Shah said afterwards: 'With the sisters controlling everything, once they accepted, everything went smoothly.'[12] The Western nations now found themselves, to their bewilderment, confronted with a cartel, not of companies, but of sovereign states.

NOTES

1. The original five had been gradually augmented, and by 1973 the members were: Algeria, Ecuador, Gabon (associate), Indonesia, Iran, Iraq, Kuwait, Libya, Nigeria, Qatar, Saudi Arabia, United Arab Emirates, Venezuela.

2. Interview with author, December, 1974.

3 See Chapter 1, pp. 32–33.

4. Multinational Hearings: Part 7, pp. 546–7.

5. Interview with author, February, 1975.

6. Multinational Hearings: Part 7, p. 517.

7. Ibid: p. 513.

8. U.S. Senate: Permanent Subcommittee on Investigations, April 22, 1974.

9. Interview with author, December 1974.

10. Interview with author.

11. *Middle East Economic Survey*: December 28, 1973; *Petroleum Intelligence Weekly*: December 31, 1973.

12. Interview with author: February, 1974.

13

The Reckoning

It is time we began the process of demystifying the inner sanctum of this most secret of industries.

Senator Church, December 1973

As soon as the embargo began, in October 1973, the seven sisters were compelled, at the risk of forfeiting their concessions, to be the instruments of the world-wide cutback in oil. They had to allocate their oil in a way that would not appear to defy the Arabs' boycott, yet would satisfy their customers throughout the world. And the American companies had to enforce an embargo of their own home country. The angry question came up ferociously: where are the oil companies' true loyalties?

To put it another way, it was an abrupt test of the companies' multi-nationality; could they, in this crisis, continue to be all things to all countries? It was not a test the companies wished to face; they preferred not 'to play god'. In the previous months some of them had made it clear to their governments that it should be their job, not the companies', to decide how the oil should be shared. But the consuming governments had no wish to face up to such divisive responsibility. There had been many discussions within the OECD about forming a common front or an emergency committee to share out supplies—particularly since the Libya fiasco in 1971. But the consuming governments could not agree. The Japanese, being the most dependent, wanted the sharing to be on the basis of overall needs of the economy; the United States, having the most internal production, wanted it on the basis of imports.[1]

Faced with a united OPEC, the consuming governments were thoroughly disunited, and quite unable to agree on a basis of rationing (rather as OPEC had originally been indecisive about pro-rationing). The sisters were thus landed with the job of serving as a kind of temporary world government, for four or five months, to re-allocate the world's oil. Or as one of their executives put it: 'We became the world's slaves, beaten, abused by all and loved by none.'

The companies were very vulnerable to pressures from every side, but specially from the producing governments who were more effectively organised and, more important to their future, able to guarantee or hold back their long-term supplies of crude oil. The companies were now seen by their critics, more than ever, as hostages of foreign powers. The concept of the companies as the instruments of their countries' foreign policy had been rapidly turned upside down.

THE SHARE-OUT

The companies were able to perform this controversial task through the intricate computer systems by which they regulated the movements of their tankers and cargoes throughout the world. Exxon, of course, had the most elaborate system, on the 25th floor of its Manhattan skyscraper,[2] with the rows of TV screens recording in green letters the movements of the five hundred Exxon tankers between sixty-five different countries. Its masterful computerised system called Logics linked its headquarters from New York, Houston and Tokyo. Now, overnight, Logics faced its ultimate test of rationality: to work out how to supply and cut back every country equally.

As the embargo descended, the staff worked late into the night and into weekends, calculating the effects of the Arab cutback, diverting the movements of the tankers, trying both to obey Arab instructions and to fulfil long-term contracts. 'It became clear to us that it wasn't a business operation, but a political one', as one executive recalled; 'that's why we wanted to stay out of decisions'. Exxon's regional presidents in Europe, Asia or Latin America were watching their headquarters like hawks and bitterly complaining after a discrimination against their part of the world. Sometimes they flew into New York to make their complaints to the board. 'Didn't you feel', I asked one of the New York managers, 'that you were ruling the world?' 'No, the world was ruling us'. It was a lesson in the burdens of a multinational corporation—in the suspicions, conflicts and sensitivities of their nation-clients.

The American companies' most awkward problem was in supplying their own country, as the main target of the Arab embargo. Before the embargo America was importing 1.2 million barrels of Arab oil a day; by February it was down to 18,000 barrels—

a drop of 98 percent. This cut meant a drop in total U.S. oil supplies of 7.4 percent. In theory, the oil companies should have been able to take oil instead from non-Arab sources, particularly from Venezuela and Iran, which had not joined the boycott. But the imports from Iran went up only from 200,000 to 400,000 barrels a day—still a fraction of the total Iranian production.[3] Exxon and the other companies dared not favour the United States at the expense of other countries, and the United States lost a rather greater proportion of its total oil supply than the other countries.

In Europe, the sisters faced a much more perilous political scene. Each country, being desperately dependent on Arab oil, was insisting on having its maximum supplies. The abortive earlier discussions within OECD had only revealed the extreme differences of view. The Embargo, when it struck, provoked the most embarrassing show of European disunity since the Common Market began. The two governments which behaved worst, in the view of the companies, were Britain and France, who were both, having moved towards a more pro-Arab foreign policy, determined to get their reward in the form of undiminished oil. Neither had much sympathy for the predicament of Holland, which alone among the Europeans had been embargoed as punishment for her pro-Israel policies.

Shell, being half-British, half-Dutch, was caught between the two governments. And the Dutch were determined to get their oil—not least because Rotterdam was the centre of the European oil market. The Shell board insisted that they must apply the principle of 'equal misery'; to do otherwise would make Shell 'the arbiter of the fortunes of nations'. They thus began a massive reallocation of oil to provide Holland with substitutes for Arab oil, trebling deliveries from Nigeria, and doubling them from Iran, so that Holland experienced no shortage beyond the average cut-back for all countries.[4] Shell and the other companies had, in effect, protected Holland from the Arabs' pressure. In the words of Professor Stobaugh of Harvard, who later analysed the results in detail, in Holland the companies 'lived up to the letter of the embargo, but frustrated the intent'.[5] 'We saved the world from being dictated to', one of the chief executives boasted to me: 'it was intolerable that small countries should change the foreign policy of big ones.'

In Britain the Prime Minister Ted Heath—in the last months

of his Tory government—had been determined not to suffer from the embargo, the more so when faced with a strike by the coal-miners. On October 21 he went to the lengths of summoning to his country-house, Chequers, Frank McFadzean of Shell and Sir Eric Drake of BP. A stormy interview followed, unreported at the time. Heath insisted that the companies must not cut supplies to Britain. The oilmen insisted that, if they did not treat all their foreign customers fairly, they would risk expropriation. With Sir Eric, Heath was specially angry; he reminded him that BP was half-owned by the government. But Drake, having taken legal advice, said that he could only obey if he were told which other countries should suffer, and that he must have his instructions in writing. At this point Heath retreated: 'It was the first showdown with the government that BP have had', Drake told me afterwards.

Shell also received a battering from Lord Carrington, the newly-appointed Minister for Energy, who was supported by Lord Rothschild, the head of the 'Think Tank', who had previously been a Shell man, but who now argued vehemently against his former employers. Shell still insisted that they would not discriminate. 'If you have to make choices', said Gerry Wagner, the Dutch chairman, on Dutch television later, 'it is a public responsibility, in this case on an international scale, which we, the oil companies, cannot carry.' That brought the issue into the open, and the London *Evening Standard* carried a headline: 'The Dutch are getting British oil'. The next day Shell publicly defended their attitude: Geoffrey Chandler, their head of government relations in London, was interviewed by the BBC on December 5 and insisted: 'It would be suicide for an international group of companies to take any other course . . . in the absence of international agreement the companies are left holding the baby.'

At the beginning of the embargo, the allocation problems were not too acute. Europe could still be supplied with Arab oil, and the United States could still receive Venezuelan oil. Had the cutbacks continued and multiplied, the companies would have faced an agonising question: should they let the Europeans go seriously short, thus incurring intense political resentment, or should they begin to supply Europe from Venezuela, at the expense of their United States markets? But the Arab producers let them off the hook. By the end of the year they had relaxed their restrictions on Europe, and by March 1974 the embargo had been lifted from the United States.

The companies were not slow to point out that they had done a job that only corporations like them could perform. Chandler later insisted that this fair distribution justified the existence of the multinationals: 'if the supply had been in the hands of minor non-integrated and state companies [he maintained in a Presidential address to the Institute of Petroleum] we would have seen acute contention between consumers and direct confrontation between consumers and producers.'

But it also appeared that, if it had not been for the domination of these seven companies, the embargo would have been far more difficult, perhaps impossible, for the Arab countries to enforce. 'It is clear' said Professor Stobaugh (referring to Saudi Arabia) 'that it is easier to have an embargo when you have only four companies to deal with than if you have many companies to deal with. If they were selling oil to a hundred companies and all the tankers were independently owned and each tanker bought oil, it would certainly have been harder for the Saudi Arabians to enforce the embargo.'[6] The whole machinery of the sisters, which had been so effective for the consumers in the sixties, holding the balance between Saudi Arabia, Iran and Kuwait, had now been turned right around: it now allocated the scarce oil to the consumers, obeying the edicts of the producers.

But this five months' balancing act was only part of the political exposure of the sisters. The smell of oil was everywhere, and in the United States, all the enemies of the companies were now waiting to pounce.

THE PRICE

The American public was so shocked by the embargo that they took some time to realise the greater shock of the price and the cartel. The ordinary motorist, accustomed to unlimited fuel for the past three decades, was astonished to find that he was dependent on Arab oil, on 'six Sheikhs and a Shah'. There had been a temporary shortage of gasoline the previous winter, due to a lack of refining capacity, which had already caused a political row; but this new shortage resulted in far wider impact—particularly on the East coast where imports were highest. By November the filling-stations were beginning to shut down, with signs saying 'Sorry, No Gas'. Cars had to drive for miles to find gasoline, and then to line up for an hour or more, with their engines running, using up more gasoline as they waited.

People immediately suspected that the companies had deliberately manipulated the crisis, and had joined the side of the Arab cartel to put up the price. The advertising and the giant signs of the companies made them all the more exposed to attack: Exxon had only recently changed its name from Esso, spending $100 million on the new signs and advertising, but everyone knew it was the old Standard Oil Company. 'We want you to know', boasted a series of Exxon TV commercials about their heroic explorations. But now the commercials came in between the announcements of new shortages and Congressional attacks, as if to underline Exxon's guilt.

Travelling in America that winter, I found the oilmen bewildered and hurt by the sudden surge of public indignation; they seemed to have no friends left. 'I must admit', said a Gulf man in Pittsburgh, 'that at parties I don't tell people I'm in oil'. In Houston, the company men talked like rejected missionaries: 'We've lifted the burdens off men's backs', said one passionate spokesman: 'we've put food in their bellies, we've achieved the recovery of Europe and Japan. And now Congressmen come to us and say: look, we've got to show our people that we can get tough with the companies: I'm sorry, but we've got to attack you.' The right-wing Texans were confused in their remedies, caught between a traditional Arabophilia and a desire for drastic action: one proposed to me that Israel should be invaded and pacified in order to remove the cause of Arab discontent. 'I see Arab money coming to the fore,' said another Texan oil-man 'and the companies caught up in anti-semitism.'

It was soon apparent, from opinion polls, that Americans blamed the companies more than the Arabs. As motorists waited in their cars, listening to the radio describing new shortages, they looked up angrily at the giant sign at the head of the line proclaiming Mobil, Texaco or Exxon: the symbols no longer of plenty, but of an infuriating and ill-organised shortage. If the signs had said Saudi oil or Kuwait oil, the reaction would certainly have been fiercer. As it was the companies were buffers in a new sense, protecting the Arabs from the public fury.

Just as the shortage was worst, and the winter coldest, the oil companies began to announce record profits. The figures were staggering: Exxon announced that its profits for the third quarter of 1973 were up 80 percent on the previous year: Gulf was 91 percent up. Exxon's profits for the whole year turned out to be an all-time record for any corporation, anywhere, at any time: a total of $2,500 million. It was true, as the companies hastened to point out, that 1972 had been a bad year for profits. The huge improvement for 1973 was therefore partly due to the low base of the year before, and also to the devaluation of the dollar. But it also came from the far higher value of the companies' stocks, and the higher profit from crude oil, as the price had gone up. The profit figures spotlighted the embarrassing fact that the companies were now—as Yamani had always intended—'married' to the producing countries, while they soon attracted, as we shall see, the interests of their marriage-partners.

The oil companies began frantically advertising to justify their profits: on one February day the *New York Times* carried three full-page advertisements, from Texaco, Gulf and Shell. They explained that profits had previously been much too low, and that they needed vast new investment to develop energy resources. The Chase Manhattan Bank (the old Rockefeller bank, still specialising in oil statistics) reckoned that $600 billion would be needed for energy development, which made the oil profits look puny. Shell even protested that their profits were too low. Mobil, always the most compulsive advertiser, finally seemed to give up trying to establish their credibility. They put a sad advertisement into the *New York Times* headed 'Musings of an Oil Person', which caused a small stir in the industry. 'Wonder if oil company advertising isn't risking indecent exposure these days,' it began, and complained that in thirty seconds the TV news 'can suggest enough wrong doing that a year of full-page explanations by us won't set straight.'

Mobil insisted that the consumer had to pay for new development: 'we're recycling the money he pays at the pump right back into oil-finding offshore, Alaska, anywhere.' But Mobil's protestations were not very convincing; only a few months later they announced that they had made a bid of $500 million for the Montgomery Ward chain of stores—suggesting that they were recycling their profits to get *out* of the energy business. Nor did Gulf add to the public confidence by making an abortive bid, in the middle of the crisis, for the famous circus company, Ringling Brothers, Barnum and Bailey. It hardly fitted in with the image of arduous and complex technology.

UTOPIA UNVISITED

During that frosty winter, it seemed for a time that the United States would at last be forced to wean itself away from its dependence on oil. The reign of Big Car seemed to have abruptly ended; in Detroit in the height of the crisis, there was a pervasive gloom about the prospects for the 'gas-guzzlers' on which the city's industry had so long depended. While sales of Volkswagens and Toyotas were booming, the American companies were hectically re-tooling to make small cars.

The value of real-estate in city centres moved up while houses in the outer-suburbs waited unsold. Housewives and commuters formed car-pools, helped by computerised systems. In Manhattan, theatres and cinemas reported record business, as the city-dwellers took more to their feet, and less to their cars. Planners became again concerned with mass-transit, looking towards Canada, Europe, and even to Moscow for models of subway systems. The whole style of modern architecture, with glass skyscrapers exposed to the elements, was called into question.

So long as the embargo was enforced, until March 1974, the trend continued. In the first quarter of 1974[7] consumption of gasoline was down 7·7 percent over the previous year, instead of the expected normal *increase* of around 7 percent. But once the shortage was over, consumption shot up again, even though the price of gasoline had risen by 40 percent over the year. Big cars came back into vogue; cars drove faster; road deaths went up. In August a bill to spend $20 billion dollars on reviving mass transit in cities was cut down by President Ford to $11 billion; and soon afterwards Ford, speaking in Chicago, clearly identified himself with defending the auto industry: 'I'm a Michigander, and my name is Ford.'

Far from improving the American environment, the energy crisis was now further wrecking it: President Nixon had relaxed the pollution laws, allowing sulphureous fuel to be burnt for heating oil; permitted more strip-mining for coal; and allowed the building of the Alaskan pipeline, against the objections of the environmentalists. In the crisis atmosphere, the ecologists carried little weight.

The Utopian vision quickly dissolved; and the crisis only served to underline the dependence on oil, without providing any alternative. As the British writers Davenport and Cooke had said in

1924: 'Does not the American partly live in oil? Certainly he cannot move without it.'[8]

On conservation the oil companies were hopelessly ambivalent. They all now theoretically supported it, preaching the virtues of turning down heating and driving slowly, and many oilmen took the view that only by conservation could the stronghold of OPEC be weakened. But the whole industry had been built up on encouraging consumption, and luring customers away from buses and railways towards the extravagance of big cars and highways. The companies had made no attempt to persuade the auto companies to prepare for a shortage by making smaller cars ('I think we did not do that, and I think we should have', admitted David Bonner of Gulf in February 1974).[9] And the oilmen were schizophrenic about increasing domestic taxation on gasoline or cars, which was the most obvious way of reducing consumption. Many of them, I found, privately agreed with the need for a higher tax, both to conserve fuel and to provide funds for alternative transport. But publicly, none dared espouse a policy that would injure their profits; and the whole hectic roadside advertising continued to spur the motorists on. The Arabs were carefully watching the effects of higher prices on American habits. The continued consumption encouraged their claim that their huge price-increases were perfectly reasonable, and that for decades the price of oil had been far too low.

THE ATTACK

In Washington the oil companies were receiving a political battering unparalleled since the Nazi scandals of 1940. Now, as then, they were accused of being traitors, supporting foreign powers at the expense of the United States. And the public anger coincided with the anger about the Watergate scandal, which was still being uncovered. As with the Teapot Dome scandal, fifty years before, oil seemed part of the general corruption of politics by big business: an attitude summed up in the neat slogan 'IMPEACH NIXXON'. Senator Ervin's Committee, enquiring into the secret campaign contributions, found two oil companies, Ashland and Gulf, to have broken the law; and uncovered the first clues in a long trail of evidence, showing how oil companies were able still to maintain a secret network of bribery. Claude Wild, Gulf's Vice-President for Government Relations in Washington, testified to how he had provided $100,000 to Nixon's presidential campaign, laundered through a subsidiary in the Bahamas; and also $10,000 to Senator Jackson and $15,000 to Congressman Wilbur Mills.[10]

With this combination of scandals, the American sisters were under fire from all sides. The most opportunist attack came from 'Scoop' Jackson, Democrat senator from the State of Washington, who was now again revealing his Presidential ambitions, and who stood to gain on several fronts. He was a steadfast champion of Israel, and he had strong support from the Labour unions. He knew a great deal about the oil companies and their history, and had close friends among them (witness Gulf's gift of $10,000). But like Harold Ickes thirty years before, he liked to play the game on both sides.

As chairman of the Permanent Subcommittee on Investigations, Jackson sent out questionnaires to the seven biggest American oil companies, who included all the seven sisters except BP, and also the Standard Oil Company of Indiana (Amoco). Jackson then summoned the supply experts of each company to hearings in Washington. Only one of the seven, Shell Oil, perceived the political danger, and sent their chief executive, Harry Bridges. The rest were technical men and they were thus sitting ducks for the Senator. They were lined up in the front row of the hearing room: Roy Baze of Exxon, David Bonner of Gulf, Harry Bridges of Shell, Annon Card of Texaco, R. H. Leet of Amoco, Allen Murray of Mobil, T. M. Powell of Socal. Together they stood up to raise their hands and take the oath in front of Jackson, to provide a front page picture of the sisters in the dock.

Jackson made the most of it, opening with a succession of menacing questions, staring at the TV cameras with his blank eyes and unmoving jowl. 'The American people want to know whether major oil companies are sitting on shut-in wells and hoarding production in hidden tanks and at abandoned service stations. The American people want to know why oil companies are making soaring profits . . . The American people want to know if this so-called energy crisis is only a pretext, a cover to eliminate the major source of price competition . . .' He was soon able to humiliate the company men. Exxon had foolishly insisted that many questions including how many filling stations they owned, involved 'proprietary information'. Jackson lashed back: 'I am just frankly flabbergasted', he repeated three times, until the unhappy Exxon expert, Roy Baze, eventually revealed the number (9,145). When Jackson asked for the company profits for the first nine months of

1973, Baze could not give the answer, and Jackson himself picked up the telephone to ask the stockbrokers.

Jackson's hearings eventually fizzled out, with the questions never properly pursued; he knew well enough that the problem was more complicated and he now turned to attacking the 'obscene profits' of the oil industry. The oilmen were frustrated and bitter in the face of Jackson's onslaught. David Bonner, the President of Gulf in the U.S., retreated back to his stronghold in Houston to give an angry press conference: 'It made me feel something like I was at a criminal trial.'

Exxon saw this and other attacks as threatening the basis of American democracy. At the annual meeting of Exxon at Los Angeles in May, Ken Jamieson stood up in front of a giant luminous Exxon sign and explained that Exxon's profits were not really high enough, for the company were planning to spend $16 billion in four years to develop new sources of energy. He revealed that Exxon were pressing for a massive educational campaign to make people understand the true facts about oil. Soon afterwards, Michael Wright, the chairman of Exxon U.S.A., produced a pamphlet called 'The Assault on Free Enterprise.' 'Let there be no mistake,' Wright warned, 'an attack is being mounted on the private enterprise system in the U.S. The life of that system is at stake.'

Scores of senators and congressmen rushed to propose new legislation to control the oil companies; according to one count, there were almost 800 bills concerned with the energy crisis. Jackson proposed a bill to enforce the 'Federal chartering' of the international companies, allowing for the appointment of a government nominee on each board, to ensure that they acted in the public interest: a kind of American equivalent of the old BP solution. Senator Adlai Stevenson III proposed a more drastic reform, that the Federal government should form its own oil and gas corporation, called FOGGO, which would have priority in bidding for new offshore concessions, and which could provide a kind of yardstick for the government to check on the profits of other oil companies.

Jackson also put forward a bill to roll back oil prices, on the grounds that the industry's profits were excessive. His charge and others led to an important argument with the *Wall Street Journal*. The *Journal* maintained in their editorials that big profits were essential to further exploration, and that oil prices should be

allowed to rise in the free market to the level where they would
reduce demand and stimulate extra production: 'In our view,
elasticity of demand is the key to the immediate problem.' Jackson
insisted that higher oil prices would not, at least in the short run,
produce extra supplies.[11] It was a variant of the old argument between oil economists surveying the history since Rockefeller: was
oil a special exception, an industry which was not self-adjusting?
In the event, Jackson proved more justified than the *Journal*: the
higher prices failed to stimulate much extra production, or to dis-
courage consumption.

But as many oilmen had predicted, by the time the immediate
shortage was over, the attack on the companies had lost much of
its impetus. By July 1974 only eight bills had actually become law.
These included bills to set up the new Federal Energy Agency, to
allow the building of the Alaska pipeline, to suspend pollution
laws, and to enforce 55 miles an hour on the roads. The other bills
had either been lost or defeated. The politicians returned to their
apathy, and the oilmen regained some of their confidence. In July
Tavoulareas of Mobil was asked: whatever happened to Senator
Jackson? And he answered:

> Well, let's analyse the situation. We had a crisis. People were
> disturbed. They had a right to be disturbed. I'm working in the
> oil industry, so I feel, I guess, defensive to an extent, and I say they
> wanted to find a whipping boy and blame somebody. My problem
> is I don't see any additional barrels of reserves coming out of all
> these investigations . . .[12]

THE HEARINGS

But the roots of the crisis went much deeper than the embargo
or last year's oil profits. And the task of untangling the longer
story fell to a Senate Committee more exhaustive than Jackson's.
In 1972 the Foreign Relations Committee had formed a subcommittee to investigate multinational corporations and their
influence on foreign policy, prompted by the attempts of the
International Telephone and Telegraph Corporation (ITT) to
bring down the Allende regime in Chile. After uncovering the
seedy background of ITT, the staff had moved on to the oil
companies, establishing a body of historical evidence comparable
in scale to the FTC report of 1952.

The chairman of the subcommittee, Senator Frank Church,
was a persistent critic of the multinationals. His bland style and

resonant voice concealed a ruthless mind: coming from Idaho, Church had a built-in suspicion of international big-business and government secrecy, which had been sharpened by his opposition to the Vietnam war. He was supported by a passionate chief counsel, Jerome Levinson, whose scepticism of big companies had been fortified by working for the Alliance for Progress in Latin America, and who was backed up by a team of relentless investigators. They were just preparing for their hearings when the oil crisis had broken on the world.

'We Americans must uncover the trail', Church said in a speech in December 1973 in Iowa, 'that led the United States into dependency on the Arab sheikhdoms for so much of its oil . . . Why did our government support and encourage the movement of the huge American-owned oil companies into the Middle East in the first place? . . . We must re-examine the premise that what's good for the oil companies is good for the United States.' Opening the hearings on Januray 30, 1974, Church made clear that his chief subject would be the seven sisters, who were all among the fifteen biggest multinational corporations in the world. 'We are dealing with corporate entities which have many of the characteristics of nations', he asserted: 'thus it should surprise no-one that, when we speak of corporate and Government relationships, the language will be that which is appropriate to dealings between sovereigns.' And Church's questioning was followed by his fellow-senators: Clifford Case, the austere pastor's son from New Jersey, chewing his spectacles through his long philosophical queries; Edmund Muskie asking with well-timed anger about the companies' tax avoidance; Charles Percy, the elegant Republican who was preparing to stand for President, determined to reassure his business friends with mellifluous tributes: but even he felt obliged to be stern with the oil companies.

The sub-committee soon produced a procession of witnesses, like a historical pageant, to justify the critical decisions over three decades. George McGhee, the Texan oilman who had helped to negotiate the fifty-fifty tax deal in the State Department in 1950; Jim Akins, sternly defending his support of the Libyans in 1970: Jack Irwin, now Ambassador to Paris, who had made the abortive visit to the Middle East in 1971. An army of representatives arrived from the seven sisters, including Howard Page and George Piercy from Exxon, Bill Tavoulareas from Mobil, Otto Miller from Socal. And the Senators subpoenaed and published a succession of past

secret documents, including the Iranian producers' agreement of
1954, the Libyan 'safety-net' agreement of 1971, and (after a bitter
fight with the State Department) the cables exchanged before the
Teheran Agreement of 1971.

The star defence witness was Jack McCloy, with his big head
and twinkling eyes: he was still representing all seven sisters, and
he had been privy to most of the key decisions since the Second
World War. He maintained that the government's interventions
had achieved both a bulwark against communism and a guarantee
of cheap oil for the West. 'The United States', he protested in a letter to Senator Church, 'has benefited perhaps more than any other
nation from the presence and activities of its companies in the
Middle East.' The current crisis, he said, had not resulted from the
companies' cartel, but quite simply from the October war. Church
retorted that the roots of the crisis went back to the 'singular
dominance of the majors': if the United States had not encouraged
such concentration, the country would find itself in less dire straits.

The story that slowly emerged, from the mountain of memos
and testimony, was not one of crude corporate clout, like the story
of ITT in Chile. It was a more intricate and fascinating tale of the
interplay of government and companies, with a gaping void of
abdication and evasion in the middle. It became clear that the
State Department, after helping to safeguard Aramco and the
Iranian Consortium in the 'fifties, had virtually delegated its responsibility for oil supplies: partly deliberately, because of the
embarrassment of the Israeli question; partly through apathy, for
the oil seemed to be flowing so freely, and the companies claimed
to be able to look after the problem by themselves. As George
Piercy of Exxon explained: 'The fact is that up until 1970–71,
there was little need for the Government's active involvement or
intervention. Oil was flowing in ever-growing quantities at low
prices...'[13]

But the companies, left to themselves, were preoccupied by
their short-term profits, and quite unable to reach any long-term
accommodation with the producers. After the formation of
OPEC they became increasingly dependent on OPEC oil, encouraged by the foreign tax credits, and less able to stand together
with independents. When the Libyan crisis broke in 1970 neither
they nor the government had any serious counter-strategy. Once
OPEC began to realise its strength, with the shortage after 1971,

the companies realised their position was weakening, but they never made use of the precious time.

There was now hardly anyone left in Washington who grasped the magnitude of the problem. The experts had thinned out, as McCloy explained, because the country was 'living in a sort of fool's paradise'.[14] There was no energy policy, and no machinery for making one, and the big companies in the meantime were preoccupied with hanging on to their profitable concessions. They were either unaware or unworried that they were becoming the hostages of the producers; and that their cartel was being taken over, and turned against them—until in late 1973 they suddenly found that (as an Aramco representative vividly admitted) they had no leverage on prices at all.

The whole debate about oil policy had been kept out of sight of the public and Congress. 'In a sense this is the overriding lesson of the petroleum crisis,' said the subsequent report of the Church subcommittee: 'in a democracy, important questions of policy with respect to a vital commodity like oil, the life blood of an industrial society, cannot be left to private companies acting in accord with private interests and a closed circle of government officials.'[15]

GOVERNMENTS AND COMPANIES

From the first moment of the embargo it was once again evident to Washington that oil was too important to be left to oilmen. The old notion of oil companies as the instruments of foreign policy had been turned on its head, as the companies appeared to be making foreign policy, and the buffers had clearly broken down. The White House and the State Department were confronted with the problems they had so often sought to avoid.

The first attempts at an energy policy were more apparent than real. In November 1973 President Nixon had made a stirring television speech, proposing his Project Independence: the United States must become self-sufficient in energy by 1980, by increasing its drilling and use of other fuels, and by massive spending on nuclear research—a new kind of Manhattan project. It was certainly in keeping with all the old optimism of the oil pioneers, that just when one source of oil is drying up another one will appear somewhere else. But now, at least within the United States, there was little justification for such confidence. By the end of the year Project Independence was a joke, and self-sufficiency by 1980 inconceivable; by the end of 1974 the United States was more dependent than ever on Arab oil.

The Defence Secretary, James Schlesinger, began hinting that the embargo might lead the government to use force in the Middle East; and hints continued to be dropped from the Pentagon. But there was little sign that the hints were taken seriously in OPEC. In February 1974 Dr. Kissinger convened a conference of thirteen major consuming nations in Washington, to try to establish a common energy policy; but the conference only served to reveal publicly what was already obvious privately, that there were major differences between the Americans and the Europeans, who were too thoroughly dependent on Middle East oil to want a confrontation with the Arabs. The chief outcome of the conference was a theatrical attack on American policy by the French delegate, Michel Jobert; the British and Germans appeared to take the American side, but soon afterwards, to Kissinger's anger, the French summoned another conference of Europeans only to meet with the Arabs.

The European nations at the peak of the energy panic were now each desperately trying to secure their own sources of oil, by-passing the oil companies, to make separate deals with the Arabs. The British Conservative government had sent out a special emissary, Lord Aldington, to Saudi Arabia to secure oil supplies, to the annoyance of the oil companies: 'waiting to kiss the hem', as one Shell director put it, 'of any passing galabea.' The French sent out a special mission to Saudi Arabia, seeing a chance to break into the all-American preserve, the *chasse gardée*, which they had been trying to enter for the past quarter-century. The fact that Saudi Arabia, the critical source of oil for Europe, was in the hands of an all-American consortium, was now more than ever exasperating to the Europeans, who saw their oil dependent on the vagaries of American companies. Kissinger retorted by attacking the bilateral 'beggar my neighbour' policies of the Europeans, and insisting that only by working together could the West be effective.

But in the meantime the Americans, too, were busily fostering their own bilateral relationships; and in June Prince Fahd and Sheikh Yamani from Saudi Arabia were received in great style in Washington, to prepare a new agreement for mass American technical and military aid, to cement the most special of special relationships.

With the Shah, too, the Americans were determined to strengthen both their own bonds and to make profitable arms deals: in 1974 Iran bought more than $4 billion worth of arms from the United States. The chaos of Western rivalries now revealed the exact opposite of the position in the early years of OPEC: now it was the Arabs who were united and the West divided.

Out of this chaos there gradually emerged a new instrument of Western collaboration, which was Kissinger's special brain-child, the International Energy Agency, which was to be the instrument of Western oil collaboration. Before long all the major consuming nations had joined it, except France—who was still determined not to antagonise the Arabs. By November 1974 the IEA had agreed on a plan for mutual assistance, to ensure that no member would be victimised in a future embargo or other sanctions; but there was great disagreement as to the wider aims of the agency. Dr. Kissinger saw it essentially, though he did not publicly say so, as a counter-cartel, to confront OPEC and break it; and his abrasive assistant, Thomas Enders, later (April 1975) openly stated that his aim was to break the OPEC cartel. But the European members had no wish for such a confrontation, and preferred to meet with OPEC members on a much less provocative basis.

The British government's relations with the oil companies had now taken a curious new turn, due to the discoveries in the North Sea, whose estimated reserves were steadily increasing. By the end of 1974 ministers were privately foreseeing large exports of oil by the 'eighties, so that the quadrupled oil price might in the end prove the salvation for Britain's balance of payments. The exploration, however, was now so expensive that it was only economic with a fairly high price (if the Arabs had *not* at least doubled the oil price the future of the North Sea would have been much more problematical). Thus, although Britain faced huge oil import costs for the next five years, her long-term interests were coming much closer to OPEC's, and the last thing the government wanted was a permanent drop in the oil-price.

But the North Sea and the energy crisis made the question of controlling the oil companies much more urgent. The parliamentary report of 1973, revealing the past generosity in handing out concessions and giving tax relief, had made politicians on both sides more wary. Both Conservatives and Socialists considered taking closer control over BP, now more than ever the Frankenstein monster, through more serious government representation;[16] and government relations with BP were icy. But the Foreign Office insisted that intervention would damage BP's relations in other countries, particularly in Alaska (the former head of the Foreign Office, Sir Denis Greenhill, was now a director of BP). Instead, the notion began to take root that the British government, like the Arabs, should engage in 'participation'. Once oil was discovered, the British felt the same need to have control over their own resources, and this applied particularly to the Scots, in whose waters most discoveries were made, and who were talking menacingly about the exploitation of their own oil and threatening secession. Ted Heath was already coming round to the idea of participation before the general election of February 1974: and one of the last acts of the Tory Minister for Oil, Patrick Jenkin, was to recommend it. The arrival of the Labour government and the energy crisis gave a new fillip to participation, and Lord Balogh, the practised critic of the companies, who came back into government was convinced that only participation could extract the true facts from the companies and safeguard British oil—particularly in the context of the Common Market.

The British government therefore proposed to buy 51 percent of the shares of each of the North Sea concessions, and to form its own British National Oil Corporation, BNOC, based on Glasgow, to look after its interests. 'No other government outside the United States', said Eric Varley, the Minister for Energy, 'has thought it wise to be completely dependent on the oil companies.'[17] The oil majors were furious, particularly BP, to whom the proposal was a clear indication of distrust. Sir Eric Drake had agreed to barbed negotiations with the Labour energy minister Eric Varley, only because, as he explained, he felt obliged to discuss any governmental proposal, 'even if it was to put a Hottentot on the board.'[18] But participation was really an updating of Churchill's old insistence that the government 'must become the owners, or at any rate the controllers' of their own oil supply: extending from 51 percent of a company to 51 percent of a sea. The American oilmen now bracketed the British with the Norwegians as being the 'blue-eyed Arabs' and the 'Sheikhs of the North': but the State Department made clear to the British government that they would not interfere.

It was certainly ironic that the British government, having

protested against participation so vigorously in the Middle East, and having denied it to the Iraqis over the past fifty years, should so quickly insist on it for themselves. But it was characteristic of oil that once found in anyone's territory, it made the world look very different.

TRUSTBUSTERS INTERNATIONAL

While the Western governments were developing new ways of controlling the oil companies, the trustbusters were trying once again to separate them, or break them up. This time they were at work not just in Washington but in several Western capitals. The global scandal seemed at last to be provoking a global response, as Rockefeller's scandal had provoked a federal response a century before.

In Washington the new anti-trust chief Thomas Kauper was less obtrusive and more detached than his predecessor McLaren, also quietly persistent. With the new solidarity of the OPEC cartel, however, the trustbusters faced a more difficult problem. The sisters had continued to get anti-trust permission for joint-bargaining with OPEC up to the last negotiation in Vienna in October 1973 on the grounds that this was in the consumers' interests. But the new kind of bargaining was not necessarily concerned with keeping prices down. For the companies were now more concerned to keep their access to the 'buyback' oil; and the negotiations with the Saudi Arabian government, for instance, showed some signs of being in restraint of trade. But the task was now much harder, for the participation agreements had bound the companies to the producing governments, and the basic cartel was one of sovereign states. As Kauper put it in June 1974: 'you are also dealing with another party to the transaction, a party which is a sovereign nation, and that makes questions of anti-trust relief in the domestic courts of the United States rather difficult.'[19]

The political thrust behind the anti-trust movements was now again very forceful, not only against the oil companies, but against all giant corporations. With the growth of the conglomerates and multinationals, individualism was more than ever threatened, and the 'Age of Combinations' which Rockefeller had proclaimed, had arrived on a global scale. And because it was global, the anti-trust problem was much more difficult. This is how a top anti-trust official put it to me in June 1974:

There have been two major changes, first OPEC are now owners of oil; second, the negotiations are about control over the crude, not about pricing. And at the same time there's been a growing concern about the question of size as such, not so much for sophisticated economic reasons, as for political and social reasons. People feel they have lost their say over their destinies in their communities, because of the giant companies: conglomerates like ITT added to the worry, but the chief concern is just with size.

The difficulty for us is to establish exactly what is objectionable legally: people have an uneasy feeling, but no-one is quite sure why. It may be like the 1880's when there was a powerful popular feeling against the trusts, like Standard Oil; but without a very effective intellectual argument.

And the competition with foreigners always complicates the question. When a foreign company comes in, there's pressure to ease controls at home. The argument that we need big companies abroad to compete with the Japanese is very plausible; but of course you can't have big companies abroad and small companies at home. Anti-trust has become much more an international problem. And it is now often hard to know the genuine nationality of a big company, as it was with the states of the Union in 1880. What nationality is Aramco now? There's some doubt whether it's possible for us to break it up, because it's now part of a sovereign power.

As the groundswell of indignation against the oil companies spread, it set in motion waves of anti-trust cases all over the industrialised world. In May 1974 Japan's anti-monopoly organisation, the Fair Trade Commission, brought its first major case against price-fixing: it charged twelve oil companies including Shell (but no other sisters) with conspiring to fix prices since 1973.[20] But the case would take years to be tried. In West Germany, the cartel office—perhaps the most effective anti-trust body— began an investigation into the price rises in 1973, accusing the companies of acting in unison to exploit the crisis. The German activity quickly reacted back in Washington: 'Why is it that the German cartel office has to be in the fore-front?' Jerome Levinson, the chief counsel of the multinational subcommittee, asked Thomas Kauper. Kauper explained that the Germans could act on the basis of 'abuse of economic power', in circumstances where he could not act. But the German cartel office had concluded that there was not much any individual country, at least of the size of West Germany, could do about it.[21]

In France the repercussions were specially rough, for the energy crisis appeared to be a humiliating defeat for French government policy which had always been more defined and more interventionist than the rest of the West. The French National Assembly in June 1974 appointed a commission of deputies to examine the oil companies operating in France, and their relationships with the State, and five months later they published an outspoken report, written by the conservative deputy Julien Schvartz. Since the French had lost their unique access to Algerian oil in 1970, he complained, they had become still more dependent on cheap Middle East oil, just when OPEC was becoming more effective: 'in this matter technocracy had taken over from politics'. The French policies towards their own companies were now thoroughly muddled: CFP, the French equivalent of BP, had become equally uncontrollable and virtually a multinational company like the others. French oil policy had become merely the result of struggles between different powers, in which the public interest was forgotten. There should henceforth be an autonomous organisation for controlling French oil policy, concluded the Report, accountable to the Assembly, and the French companies must be answerable to the public.[22] In the meantime the public anger had been further stirred up by the public prosecutor in Marseilles, who in February 1974 indicted the chief executives of 43 companies, including the French heads of Exxon, Shell, Mobil and BP, on charges of discriminating against independent distributors.

The European Community also reacted to the oil companies with unaccustomed firmness. They commissioned a full report on the oil companies' behaviour during the crisis, and in November 1974 the Commissioner for Competition, Albert Borschette, sent a complaint to each of the seven sisters in Holland accusing them of having abused their dominant position during the shortage, by cutting back supplies to the smaller Dutch oil companies.[23] This broadside from Brussels provoked very different reactions from the Europeans. While France and Germany were in a radical mood against the sisters, Britain and Holland were much more protective—both because they were the home ground of two of them, and because they both now had oil and gas of their own. The British oilmen's attitude to the anti-trust suit was fairly contemptuous: 'The independents make their money from surpluses,' said one Shell director: 'to save them is like lending a bed to your wife's lover.' But that was what worried the trustbusters everywhere; that the shortage strengthened the hand of the giants.

Anti-trust actions had certainly gained momentum from the oil crisis. But there remained a deep division of attitudes, according to interpretations of the whole history of the companies. Was the crisis for the West so serious that they could not afford the democratic luxury of anti-trust—as the State Department had decided in encouraging the Iranian consortium twenty years before? Or were the companies now really part of the cartel on the other side, underpinning the new cartel of sovereign states? And if so, how could any committee of lawyers be effective against *that* combination? Against that line-up, was the only ultimate anti-trust action—as some hawks were beginning to suggest—a war?

NOTES

1. See Testimony of Professor Robert B. Stobaugh (July 25, 1974) in Multinational Hearings: Part 9.

2. See Chapter 1, pp. 26–27.

3. Multinational Hearings: Part 9.

4. Paper issued by Shell Nederland BV to Dutch parliament, March 18–20, 1975.

5. Multinational Hearings: Ibid.

6. Multinational Hearings: Ibid.

7. According to statistics from a sample of States, assembled by the API.

8. See Chapter 4.

9. Interview with Morton Mintz: CBS Face the Nation, February 17, 1974.

10. Watergate Hearings. See also Chapter 9.

11. See (specially) *Wall Street Journal*, March 13, 1974.

12. CBS TV Network: July 28, 1974.

13. Multinational Report: 1975, p. 15.

14. Multinational Report: 1975, p. 15.

15. Ibid, pp. 17–18.

16. Particularly after the collapse of the Burmah Oil Company in 1975 whose 20 percent holding in BP was then taken over by the Bank of England, giving the government the choice of a more dominating influence.

17. House of Commons, April 30, 1975.

18. BBC Radio interview, May 1, 1975.

19. Testimony to Multinational Hearings: Part 9. June 5, 1974.

20. *New York Times*, May 29, 1974.

21. See Multinational Hearings, Part 9. Also *The Economist*, May 4, 1974.

22. *Sur les Sociétés Pétrolières Opérant en France*: Rapport de la Commission d'Enquête Parlementaire. Paris, November 1974.

23. *The Economist*, November 16, 1974.

14

The New Cartel

It's taken OPEC fifteen years to put together an absolutely magnificent organisation that succeeding in getting world oil prices up by a factor of four in a month's time. They'd have to be utterly mad to do anything as stupid as go out and compete with each other to drive down the price of oil.

Bob Dorsey, Chairman of Gulf Oil, July 1974

Sovereign Nations cannot allow their policies to be dictated, or their fate decided, by artificial rigging and distortion of world commodity prices.

President Ford, September 1974

After the embargo had been lifted by the middle of 1974, the consuming countries were having to face the apparently unalterable fact that the world's oil was not controlled by a cartel of sovereign states. Their attention was focussed on the headquarters of OPEC in Vienna, which had for so long been regarded as insignificant.

No city in Europe evokes more poignantly the past glories of Western civilisation than Vienna. Along the wide ring road that circumscribes the old city, the baroque and rococo palaces loom up on either side with a melancholy grandeur. Their high domes, their ceremonial entrances and broad stairways seem all to be waiting for the ghosts of earlier eras—of Maria Theresa, of Metternich, of Haydn and Mozart. Driving round the ring road, past the baroque opera house and the tall gothic town hall, a single bleak modern block suddenly sticks out with a facade made of slabs of white lavatorial marble. Over the workaday entrance are the words TEXACO: for it houses among other people the Austrian headquarters of Texaco. But among the brass plates there is also one that announces: ORGANISATION OF PETROLEUM EXPORTING COUNTRIES (OPEC): 1st and 2nd Floors. There is nothing in the surroundings to suggest that this represents the headquarters of the biggest financial power in the history of the world.

For most of the year there is nothing exciting about the offices. On the white walls small black-and-white photographs depict scenes from the member-countries: a view of a storage tank, seen through some bushes in Iraq; a gas-oil converter in Libya; a tanker off the dead-flat coast of Saudi Arabia. Down the corridors underfurnished offices contain statisticians, economists, and cosy British or Austrian secretaries. On the floor above sits the secretary general, who changes every two years. It is not a flamboyant position: for the two critical years, 1973 and 1974, when OPEC suddenly realised its full power, the secretary was a quiet courteous Algerian. Dr. Abder Rahman Khene, who looked like, as he once was, a doctor: patient, courteous, with a neat moustache and a waistcoat, he seemed not at all excited by the sudden new balance. He saw the change, as he put it to me, as one not towards power but towards reality. The West, he believes, should be able to face willingly the reduction in its standard of living, for the betterment of the third world.

For years the meetings of OPEC had attracted scarcely more attention than other trade organisations—only a handful of specialist journalists would attend. But after the sensational turnabout of 1973, the bleak offices became suddenly the object of intense curiosity for the rest of the world. The interest had reached a new peak by the time of the annual meeting in December 1974, a year after the sensational price-rise, when the price of oil was once again to be settled. Planes flew in from Caracas, Kuwait, or Djakarta. The four principal hotels in Vienna filled up with unpronounceable names from little known countries. At the Hotel Imperial opposite the Opera, where Wagner once stayed, an elegant yellow suite was prepared for Dr. Jamshid Amouzegar, from Iran. At the Hotel Intercontinental expectancy centred round room 1141 on the top floor, which was awaiting the arrival of Sheikh Zaki Yamani.

Journalists flew in from the four corners; no longer now only the dedicated band of experts, from *Platt's Oilgram,* or *Petroleum Intelligence Weekly,* or the *Middle East Economic Survey,* but a whole pack of pundits and popularisers, grappling with the complexities of barrels and buy-back, equity and discount. Vienna was the centre of a curiosity that had never been accorded to Chancellor Kreisky or to the negotiations on Mutual and Balanced Force Reductions. For this curiosity affected every consumer in the industrialised world: what will be the new price of oil? And will the OPEC cartel hold together?

On the first morning reporters and cameras are packed into the reception room, waiting for the visiting potentates. Abdul Rahman Atiqi, the Kuwaiti oil minister, strides in behind his Groucho moustache; he has command of three million barrels of oil a day — a tenth of OPEC's production. Dr. Amouzegar follows, his humorous mouth constantly changing its shape, always good for a joke — he is backed by six million barrels. He implies gently that he does not altogether approve of three Arab states having put up the price before the meeting. Then comes the Iraqi minister, Abdul Karim, pale and inscrutable; then the Venezuelan, Dr. Hernandez, urbane but uncommunicative. They assemble in the conference room sitting squeezed round two long tables with their flags in front of them, and their advisers behind. At the head is M. M'Bouy-Boutzit, the delegate from Gabon. Round the tables are mixtures of face and colours, black, brown and white, with no imaginable common cause except oil.

Long after all the others have assembled, a large Mercedes draws up, and out steps the elegant figure of Zaki Yamani, immaculate in a long tailored black raincoat. The microphones like snake's heads push into his face, and the questions spurt out about the price of oil. He weaves his way through the cars, almost caressing his words, 'it has to go down, it has to go down', and drives off in another Mercedes. The tape recorders re-echo his phrases 'it has to go down, it has to go down'. Everyone seems to want it to go down, but still it goes up. Is that, I ask myself, the mystery-story of OPEC? It is like the *Murder on the Orient Express*. Each suspect points to one of the others: but really, they all did it together.

The next morning, after two bomb scares, OPEC moves the meeting to the Imperial Hotel. The Wagnerian lobby with marble pillars and triumphal staircase is filled with moustachioed delegates and anxious reporters: elderly hotel guests descend the stairs to find the way blocked by a battery of cameras and arc-lights. At last in the afternoon Dr. Amouzegar comes down the stairs beaming; there is good news, oil will not cost a penny more. OPEC will simply follow the precedent of the three Arab states a month earlier, reducing the posted price while increasing taxes and royalties; and the price will not be changed for nine months. Hectic discussions follow between the journalists and experts, which yield a dozen different conclusions. But eventually a French

journalist persuades one delegate to agree that, *oui*, there was a
small *augmentation*. And the eventual consensus is that oil will
now cost the West about 3 percent more—or two billion dollars.
That figure would have caused an uproar eighteen months earlier;
now it seems so modest that there is a huge sense of relief, even
gratitude.

In the evening, the reconciliation seems complete. In the Palais
Schwarzenberg, a sumptuous pile once inhabited by Metternich's
successor, Dr. Khene gives a banquet for four hundred people.
Under the ornate ceilings, the tapestries and chandeliers, the
Arabs sip orange juice, while Western diplomats and businessmen
slip anxiously among them, downing whiskies, trying to find out
what they are doing with their money. At dinner, the Austrian
Chancellor Bruno Kreisky receives the chief delegates at his
table: Dr. Amouzegar from Iran, Belaid Abdesselam from
Algeria, with a nose like a parrot, Shettima Monguno from
Nigeria, a black aristocrat with tribal scars. The old world is
paying tribute to the new.

It was a change in the balance of power at least as dramatic as
the Congress of Vienna 160 years before, and the West had now
apparently come to terms with it. There was not much serious
talk of refusing to accept petrodollars, or of invading the oilfields.
The fact now seemed irreversible, that much of the world's
wealth had suddenly shifted to an obscure corner of the world.
Discussing it at the banquet, no one could think of a precedent
for such a transfer of wealth. Was it the Spanish Conquistadores,
bringing back gold and silver to Spain? Was it Britain in the early
nineteenth century? But both these increases were less sudden and
less extreme.

And this was not a military upheaval, like the sack of Rome or
the fall of Constantinople: the war preceding it was technically
a defeat for the Arabs. Nor was it obviously caused by any sudden
failure of the West. Certainly it had some relationship with the
decolonising process by the imperial powers over the past thirty
years. Dr. Khene, when I talked to him, like many Algerians saw
the change as part of the general awakening of the oppressed
people, and their struggle for independence. To that extent,
there was nothing a liberal anti-colonialist could object to in this
new turnabout in Vienna. It was an extension of the decolonising
process into the economic sphere, where the enemies had been not
the Western nations but the Western companies. And the sisters

still served as convenient bogeys both to unite the disparate members of OPEC, and to divide the West. When, at the Hotel Imperial, Dr. Amouzegar was asked whether the new terms of the oil-price would not hurt the majors, he looked at the journalists with a smile of surprise: 'Are you *for* the majors?'

Looking back at the slow history of the producing countries, with the advantage of hindsight, the puzzle was not so much why this victory had happened, as why it had not happened earlier. These sovereign states, with their own precious resources, had taken a long time to awaken to the need to safeguard their wealth, and to combine to confront the companies. The change had been more one of confidence, than of a tangible shift of power. What had held back OPEC had been the 'mystical powers' of the companies, as the Shah described it. As one Saudi explained, 'it was not until we realised our strength that we had our strength'.

And this education had come as much from the West itself, as from any development of Arab or Iranian nationalism. Many of the milestones in Middle East history had come from London and Washington. It was the Texas Railroad Commission which taught the Venezuelans about conservation. It was Labour's nationalisations in Britain in the late 'forties that encouraged Mossadeq in Iran in 1951; it was the FTC Report of 1952 that taught the Arabs about the cartel; it was Texaco and the University of Texas that turned Abdullah Tariki, the co-founder of OPEC, into an effective radical.

The OPEC revolution was the consequence of the old dilemma, of running any liberal empire which educates its subjects to rise up against it. But now, of course, the producing countries had no-one else to blame for their new difficulties, and they were taking over the dissensions, as well as the confidence, of the West. When in April 1975 King Feisal was assassinated by a Prince of his own family, it appeared that the Prince, had learnt some of his radicalism from the universities at Berkeley, California, and Boulder, Colorado.

RICHES AND POVERTY

The sudden vast transfer of wealth was traumatic not only for the consumers but for the producers, as they began to realise first the extent of their new billions, then the limitations to them. Oil was once again playing the game of winner takes all, which it had played in Pennsylvania or Texas, but this time in the most ironic of all settings, in desert cities ruled by puritanical rulers.

Travelling through the Middle East a year after the price-rise, I found astonishment everywhere at the discrepancy between the abundance of revenue and the difficulty of spending it for the benefit of the masses. The money seemed like water on parched earth, splashing off the top and trickling down a few cracks. From outside, the oil countries may seem like the very centre of the world's stage; but from inside, they appear more like some chaotic backstage, with improvised façades propped up by a few hectic stage-hands.

This is how the experience was described to me by the director of the Kuwait Fund, Abdullatif Al-Hamad, who found himself with over a hundred million dollars to dispense in the first year after the price increase. 'You don't realise that on October 16th we got as great a shock as you did. We thought we were pygmies facing giants. Suddenly we found that the giants were ordinary human beings; and that the Rock of Gibraltar was really papier mâché. We had been protected from the rest of the world by that beautiful but unfortunate umbrella of the oil companies. We were really awed by the responsibility; it was like coming of age. We were amazed to discover that what we said and did made a difference to people in Minnesota.'

I put it to him that he could not expect great compassion from the West: it was like Rockefeller asking for sympathy. He agreed: but the problem, he said, was how the Arabs and the West could together co-operate in investments for the benefit of the Third World. The year of shock and the year of adjustment must give way to the year of co-operation. Other leaders in Kuwait were equally bewildered by the resentment about their investment: 'I never thought I'd live to see the day'—said Ali Atiga, the Secretary of OAPEC—'when a superpower like the United States worried because Kuwait wanted to buy 10 percent of a Michigan bank'.

In Kuwait the relationship between the wealth and the way of life is specially baffling. The tiny country, unlike most of the others, has enjoyed surplus wealth for the past twenty years, yet the per capita income is still only one-sixth of that of the Europeans. While the Kuwaitis themselves have a guaranteed income of $3,000 a year, many immigrants who make up the majority appear to be living close to the breadline, in mean shacks outside the city. The stalls in the souk are still full of the traditional ware— jewellery, robes, trinkets, rolls of cheap cloth—but alongside are shops stacked with cassette-recorders, stereos, colour TV's, video-cassettes. It is impossible to tell from looking at the passers-by, in their burnooses, whether they are beggars or billionaires.

The whole townscape gives the impression of unabsorbed wealth. The cube-shaped houses are plonked on the desert, with great gaps of sand between them, like bits of Lego scattered on a sandpit. The low horizon is only broken by a few lonely towers—a Hilton, a Sheraton, an office block—and a cluster of painted water-towers, shaped like golf-tees, dominating the desert: an appropriate symbol, for water is even more precious than oil.

Thirty miles outside Kuwait City, the oil city of Ahmadi, named after the ruler who first signed the concession, evokes a quite separate world. It still evokes a prim Englishness, with its yellow letter-boxes, roundabouts and verges of marigolds, and it still carries the stamp of the two companies, BP and Gulf, which first set it up, with echoes of an intense colonial community.[1] There are still three clubs, an amateur theatrical society, and a mansion for the managing director called the White House in the middle. But there are now only 130 Englishmen and a handful of Americans left there, and almost the whole business is done by Kuwaitis. Nor is it very complicated; the oil simply flows out through the oil wells under its own pressure, has the gas extracted from it, and flows under its own pressure down to the docks and the tankers: at the cost of 7 cents a barrel.

In Saudi Arabia, with a far bigger area and population, and with a more recent surplus of wealth, the problems seem more unmanageable—and the contrast more bizarre, between future prospects and present limitations. The chief signs of wealth are still the green-roofed royal palaces, like Hollywood villas, which line the main street of Riyadh; or the shops full of circular baths and florid chandeliers. There are many signs of hopeful developments; technical colleges, brand new computerised hospitals, three-lane highways leading to nowhere. Outside Riyadh, in the middle of the desert, is a small green notice saying 'Site of the University of Riyadh'. But for the moment the problems of management and bottle-necks are intractable. The tiny Saudi elite is constantly lured away by private business or to long sojourns in London or Beirut. The Minister's offices rely heavily on immigrants, particularly Palestinians, to manage their problems; but the top jobs are kept for the Saudis. Like any conservative monarchy, the country faces the problem of building up an educated elite, without threatening the political structure.

The likelihood of this vast shapeless country of six million people ever being able to absorb the oil revenues seems very distant. But the Minister for Planning, Hisham Nazer (one of the ablest of the new Arab technocrats) assures me that his future plans can make use of all the money. Down at Aramco Frank Jungers has little doubt, as he puts it, that 'the insatiable government machine will get used to digesting the money.' In spite of the political turmoil of the preceding year, Nazer and the Saudi government look to Aramco to provide the technology not only for oil, but for all kinds of development—agriculture, roads, hospitals, ports. The four sisters are apparently more than ever the agents of a foreign government: 'Our concern,' said Jungers, 'is how to be even more closely associated with the Saudis than we are.'

Visiting Teheran, the whole urban scene suggests the basic contrast between Iran and Saudi Arabia. This is the capital of a country of 32 million people, who need all the money they can get, where the oil is running out fast. The city itself hardly looks like the centre of a booming economy: it looks like the backside of another city, with pavements collapsing into torrents of water, rows of gimcrack houses, and a few isolated skyscrapers rising out of the mud. The streets are jammed with brand new American cars and double-decker buses, caught in an endless congestion. The scene might have served as a symbol for the problem facing the whole country, that the infrastructure cannot cope with the expansion, and that the developments are held up by the bottlenecks—at the ports, on the roads, in the assembly-plants.

And behind these worries is the central problem of the planners I spoke to; that their economy is still desperately dependent on the single slippery product. When the price of oil drops, or the oil runs out, they might be left with only caviar and carpets. However much a Westerner may blame the Shah for beginning the whole wave of Western recession and inflation, in the setting of Iran the Shah's insistence on a high oil-price seems a matter of patriotic necessity.

SHAH v. SHEIKH

By 1974 the long rivalry between Iran and Saudi Arabia, which had preoccupied the oil companies through the 'sixties, was the most critical question within OPEC, and the most likely source of its disunity. The most fundamental difference between the OPEC countries was not so much between African, Asian and Latin American, as between the countries with large populations, which urgently needed money, and those with small populations who could afford to take a more relaxed view about their oil. Algeria, Nigeria and Venezuela were among the former; the Gulf Sheikhdoms most spectacular among the latter.

The contrast between the two biggest oil-producers pointed up the division. Iran with its reserves of sixty billion barrels, and Saudi Arabia with its reserves of 130 billion (and perhaps far more) with only a fifth of the population. These figures alone ensured that Iran would be much more aggressive in maintaining a high oil-price. But there was no question that a key factor was the Shah himself. Having now reigned for thirty-three years, he had seen the Middle East transformed by oil revenues. After his dependence on BP and his humiliation by Mossadeq and the CIA, he had developed into a far more confident ruler, himself now one of the Oil Kings of whom he had once been in awe.

I talked with him in February 1975 in St. Moritz, in the luxury hotel close to his chalet where he skis every winter. It is a turretted palace, with a strong flavour of the 'thirties, a setting which reinforced the earlier image of the Shah as a kind of comic-opera Ruritanian monarch, but the luxury Swiss hotel had now become a waiting-place for bankers and finance ministers seeking deals from the Shah. As I waited with a courtier in the cavernous lobby, there was a flurry at the entrance where a figure appeared in a red-white-and-blue plastic ski jacket, sleeves wrapped over the shoulders, still wet from the ski-slopes. He walked briskly through the lobby followed by a plastic procession, which I was motioned to join, marching up to a hotel-room where we were left miraculously alone. His face was hard, with sharp-looking teeth, but occasionally transformed with a quick smile.

In conversation he emerged not so much as a megalomaniac, but as a man still dazed by his own sudden success. As he recalled the anxious uncertainty of his first years as Shah, when the oil company seemed to be all-powerful, he sounded again like that uncertain young man. He was puzzled by the American hints of possible invasion of the oilfields, which had been recently emanating from Washington, and clearly worried by American public opinion. But he was still insistent that the price of oil was too low, and he would keep it up, he blandly explained, for the West's own sake: 'It would really be weak and short-sighted, and it would not help the world at large: because you will start to go again into a false euphoria of cheap oil, and forget about extracting your coal and searching for new sources of energy. It would be the biggest mistake which would be made to our civilisation, I won't be part of that mistake.'

He saw himself as essentially performing a necessary task, both for the West and the Third World: 'The West made their fantastic fortune and their lavish spending on very cheap oil, without paying attention to oil producing countries, and even doing nothing for the poor countries of the world. We have had twelve years of UNCTAD[2] now—what have they done in those twelve years for the poor countries of the world? Nothing at all.' As for the oil companies, he saw them as having no real power, as they revealed at the time of the embargo: 'The companies were the first to say: "I serve and obey the orders of the countries" '.

In this anonymous hotel room, he could have been just another athletic banker: but in Teheran where I arrived a few days later, the image was very different. The face of the Shah looked down sternly in full regalia from photographs in every government office and shop, and His Majesty was talked about with almost religious veneration. It was an awe with a very practical basis: there had recently been still more allegations of torture by the Shah's secret police, Savak; and soon afterwards he abolished all rival political parties. But the veneration had a positive side too. Not even the Shah's opponents denied that he had in one single decision in 1973 transformed the whole economy of the country. The burden of debt was lifted overnight, and instead the world queued up to borrow money.

The Shah had always been distrustful of the Saudis, but now, at the beginning of 1975 his suspicions were much more pervasive. For he and his oil minister, Dr. Amouzegar, were convinced that the Saudis had made a private deal with the Americans, by which Saudi oil would be favoured to Iranian oil, and OPEC effectively broken. Although the Shah talked contemptuously about the seven sisters, he clearly suspected that if oil consumption went down further, they might regain the power to divide and rule, particularly between Iran and Saudi Arabia. The Shah's suspicions centred not so much round King Feisal—or round his subsequent successor King Khaled—as round the upstart commoner who was rivalling him as the spokesman for Middle East oil on the world's stage, Sheikh Zaki Yamani. The rivalry between the Shah and the Sheikh[3] had become a global drama played out on TV screens.

Yamani, too, was revelling in the new publicity, touring the Western capitals to present his case. His relationship with the royal family in Riyadh was the subject of constant speculation, and there were frequent rumours that he would resign. In December 1973, as we have seen, King Feisal disagreed with him about quadrupling the price, and in September 1974 there was another internal crisis when Yamani wanted to conduct an auction of Saudi oil which, it was widely hoped in Washington, would help to bring down the oil price. But there was opposition to the auction both from within the royal circle, notably from Prince Fahd, and more importantly from other members of OPEC, led by Iran and Kuwait. They eventually persuaded Yamani to drop the plan in return for other favours. Yamani, it seemed, was losing influence; and he faced a new crisis in March 1975 when King Feisal was assassinated. The key man in the cabinet became his half-brother, Prince Fahd, who had often disagreed with Yamani. But Yamani still survived, as the man who knew more than anyone about the problems of oil.

Yamani had always appeared, even more than most Western-ised Arabs, to be caught between two worlds, in robes and burnoose one moment, and in a dark suit the next. In his office in Riyadh he has all the trappings of Arab pomp, sitting behind a two tier brown plastic desk, surrounded by chrome-and-glass furniture and plastic armchairs, holding court while anxious petitioners and oilmen come and go. He dictates to a team of male secretaries, talking in fast Arabic interspersed with a few English words—liquefaction, gathering station, *crude*. He seems to be acting all the time, rubbing his nose or fingering his little beard, or joking with a dart of his tongue, enveloped in an Oriental sense of power. Yet talking to him in English alone, he seems a complete cosmopolitan rattling off statistics for barrels and buy-back, as attuned to every nuance of Washington politics as an international lobbyist.

Yamani's duality was distrusted by both sides. To fellow-members of OPEC, and most of all to the Shah, he appeared in cahoots with Washington. But to Americans, he was always promising to drop the price of oil, while it still went up. His ambiguity was part of the predicament of his country—torn between its special dependence on the United States, and the growing pressures from fellow-Arabs and OPEC. Yamani, as he had so often warned the oil companies, could not afford to stick out against OPEC.

Yet he was confident of his long-term relations with other members of OPEC: he knew that Saudi Arabia would emerge as the key piece in the jig-saw of oil. At the moment of extreme shortage during the embargo, Iran had been able to win the day. But as the shortage receded it was the Saudis—with their far bigger reserves—who were bound to carry most weight. They were like the Texans in the 'thirties: without their support there could be no control of the market, no guarantee that there would not be a flood of cheap oil which would undermine the world market and the other producers.

Yamani was in a position to be able to tell OPEC: only if they agree to keep the price at a reasonable level would Saudi Arabia agree to control its production. He was confident that he could both control and safeguard the cartel. As he put it to me: 'Usually any cartel will break up, because the stronger members will not hold up the market to protect the weaker members. But with OPEC, the strong members do not have an interest to lower the price and sell more.'

THIRD WORLD

The members of OPEC had very different views of the world, and particularly of their relationship towards the poorer countries of the Third World. It was clear from the beginning of the price-increase that the nations which would suffer most would not be the industrialised countries, however bitterly they complained, but the developing countries in Africa and Asia. For they were dependent on oil not only for fuel but for fertilisers, and they had no lucrative exports to compensate for their higher import bill. The OPEC countries, the *nouveaux riches* of the Third World, were very vulnerable to the charge of exploiting their weaker brothers. But at the same time the success of OPEC provided a stirring example to all those countries who complained of low prices for their commodities, and a challenge to take the initiative into their own hands. The more radical members of OPEC now depicted themselves as the advance guard of a new age of economic redress.

The chief spokesman of this movement was the President of Algeria, Colonel Houari Boumédienne. His country at the top of

Africa, with its own oil but with much poverty, was well placed to be the bridge between rich and poor. Boumédienne himself, with his mixture of demagogy and well-ordered French argument, was in a good position to present the case. Four months after the price had been quadrupled, he convened a special session of the United Nations in New York to discuss the problems of raw materials, deliberately timed to coincide with Kissinger's initiatives. He delivered a marathon address in which he presented OPEC as the new leaders of the Third World. 'For the first time in history,' he declared, 'developing countries have been able to take the liberty of fixing the prices of their raw materials themselves.' The example, he said must be followed by producers of other commodities, including copper, iron ore, bauxite, rubber, coffee, cocoa and peanuts.

OPEC's success. Boumédienne explained, depended on taking control if its own resources, and no recovery was possible while the multinational corporations remained in control, 'those past masters at the art of making concessions in order to safeguard the essentials'. Nationalisation could tear down the barrier which foreign companies erected between producers and clients, and could bring the developing nations face to face with the realities of world industry. It was a stirring speech and it brought a new scent of power to the UN. In the delegates' lounge, I sensed an awakening from a long torpor: the Third World was no longer petitioning for favours, but organising to insist on them. And the participants at the conference duly approved a declaration of principles to achieve what Boumédienne grandly cal'ed 'a new international economic order'.

Western diplomats and politicians, not surprisingly, were ambivalent towards Boumédienne's initiative. On the one hand, the prospect of a succession of mini-OPEC'S, progressively cartelising the commodities on which the West depended, presented a nightmare of inflation and recession. On the other hand, the Western liberals had a well-founded sense of guilt towards the Third World, having miserably failed to meet their own targets for aid. Oil was the subject of special remorse. For the substance that was burnt up by gas-guzzling cars was also now the basis of petrochemicals, and fertilisers which could transform the agriculture of the poorer countries. The inherent wastage of burning this noble fluid had been first noted by the Russian chemist

Mendeleyev in 1872, after visiting Pennsylvania.[4] But the wastage was not the subject of major concern until the sixties, when the world's oil was being rapidly used up, at a time when its uses were constantly increasing. It was a theme that the oil producers eagerly took up; they were conscious of the West's sense of guilt. The use of oil to feed the starving millions added to the case for redistributing wealth and resources. But it was one thing for Western liberals to advocate voluntary aid and improved terms of trade: quite another to have such a redistribution forced on them.

Boumédienne's vision of OPEC concealed a good deal of hypocrisy, for the high price of oil continued to be the cause of great hardship to poorer countries, and his proposals for a new economic order were not backed by any great generosity from Algeria. But other OPEC members, in the year after the price rise, did give very large sums in aid to the developing world, and went to some pains to make new links, particularly with India.

Boumédienne's rhetoric and statesmanship helped to give a sense of moral confidence and panache to OPEC, which reached a climax at the meeting of Heads of States in Algiers in March 1975. Here, OPEC presented itself not as a grasping cartel, but as the standard bearer of the new world order, bringing a new sense of justice and redistribution of the world's wealth, and a unity of purpose among its members—symbolised by the spectacular reconciliation between Iran and Iraq.[5] And at the subsequent meeting with the consuming countries in Paris in April the OPEC delegates insisted on regarding themselves as representing the whole Third World, to discuss the whole range of problems. But the central theme of Boumédienne's original initiative, that OPEC must be followed by other 'Pecs', met with little success. OPEC members helped to finance other commodity organisations, and producers of some, notably bauxite and sugar, drastically pushed up prices. But by 1975 the prices of most raw materials were tumbling down; and OPEC was less keen to associate oil with them.

It was again clear that oil was a unique commodity, for all kinds of reasons. It was by far the most important, with a world trade of four times as much as the next most valuable commodity, copper. It could not be stockpiled or stored for long-term supplies, except under the ground where it originated. Its whole history had shown since the Pennsylvania days, an alternation between shortage and glut, which was not easily subject to the rules of the market; and this tendency was now increased by the fact that several countries with the most oil, like Saudi Arabia and Kuwait, had the least need for money.[6]

And the oil business had already been in the hands of the partial cartel of the seven sisters who provided a world marketing and allocation system, so that the formation of a producers' cartel was made very much easier. As the *Economist* commented (March 8, 1975) 'many poor primary producers would give their eye-teeth if big foreign capitalists would kindly arrange a semi-monopolistic distribution network for their products in the West, down to tied filling stations'. In the public speeches at the UN conference, the multinational corporations were the favourite scapegoat, the engines of oppression and exploitation. But in private, I noticed, the attitude of some delegates from producing countries was somewhat different. As one of them put it, 'you make the cartel, and we'll take it over'. And they looked with real envy at the unique oil cartel, which seemed to be able to defy all the laws of the market.

NOTES

1. The earlier community at Kuwait Oil Company was the subject of an unpublished book by Ian Fleming, who was commissioned by William Fraser, later the second Lord Strathalmond, to write a book about it for £5,000: but the book gave a salacious account of life in the compound, and the Sheikh would not countenance its publication. It was paid for, and never published.

2. The United Nations Conference on Trade and Development was set up in 1964 as a permanent body.

3. But the title Sheikh in Saudi Arabia is merely a courtesy title and does not, as in the neighbouring Sheikhdoms, indicate royal blood.

4. See Tugendhat and Hamilton: *Oil, the Biggest Business*, Revised edition, London, 1975, p. 120.

5. See Chapter 1.

6. 'High prices prevailing in the markets for grain, cocoa, coffee, cotton, sugar and others almost inevitably lead to an extension of production,' commented Paul Frankel after the UN Conference, re-stating his theory of oil cartels, 'which by its own weight provides for a downward correction; in today's oil set-up the opposite is the case, because of the limited absorptive capacities of some of the main producer countries for the inward flow of funds.'

15

The Seduction

I'm the Sheikh of Araby
Your love belongs to me
At night when you're asleep
Into your tent I'll creep . . ..

By the end of 1974, after the departure of President Nixon, there
were signs of firm policies in Washington to try to break the
OPEC cartel, and to bring down the price of oil. The price, said
William Simon, the Secretary of the Treasury, was exorbitant: 'I
can think of no single change,' he said in November 1974, 'that
would improve the outlook for the world economy more than a
substantial decrease in the price of oil.' The political implications
of the financial power of OPEC, said Henry Kissinger, the Secretary of State in the same month, were 'ominous and unpredictable.
Those who wield financial power would sooner or later seek to
dictate the political terms of new relationships.'

The bankers and economic advisers were convinced of the
urgent need to bring down the oil-price. The increase was reckoned to have been responsible for adding 2 percent to the rate of
inflation, and the recession was showing itself everywhere. A few
banks crashed in America and Europe, and Italy was apparently
heading for bankruptcy. The World Bank estimated that the surplus funds from oil revenues would by 1985 reach the staggering
figure of $1·2 trillion ($1,200,000,000,000). The prospect of
channeling these funds into productive investments was way
beyond the resources of any private banking system. The governments, both in Washington and Europe, were determined to cut
down oil consumption, both to reduce imports and also to encourage disunity within OPEC. For if the members had to cut back
their production, it was reckoned, there would be deadlock between them, as there had been in the 'sixties, as to who should lose
revenues: there was still little sign that they were able to ration
production between themselves.

A year after the great price-rise, at the beginning of 1975, the world supply of oil was once again well ahead of demand. The industrial countries were now at last consuming less than the previous year, partly because the oil was so much more expensive, partly because the winter was exceptionally warm, but chiefly because of the world economic recession. The signs of new plenty were evident everywhere. The storage tanks were overflowing with oil, the loading docks were half-empty, the freight-rates for tankers dropped lower and lower, the great Tapline from the Aramco oilfields to the Mediterranean was now comparatively so expensive that it was closed down. Fleets of tankers were laid up, causing disaster for speculators and the financial collapse of the oldest British oil company, Burmah Oil.

Washington observed this new trend with great expectations. At the Treasury the ebullient William Simon, who had previously been energy 'czar', had always been convinced that the free market would make itself felt in oil, as in everything else; a fall in consumption would bring prices down. He had very publicly pinned his hopes on the auction of oil planned in Saudi Arabia in September 1974, and was bitterly disappointed when it was cancelled. But he was still confident of lower prices. And Henry Kissinger was working energetically to break the solid front of OPEC. He had inspired much tougher statements from President Ford, and had pushed forward the International Energy Agency. By February 1975 the members of the IEA had hopefully agreed to cut their joint consumption by six million barrels a day, thus threatening seriously to bite into OPEC's total production.

Simon and Kissinger waited for OPEC's collapse, and watched the growing indications. To many economists, it seemed a matter of time before the cartel was broken. Milton Friedman, the high priest of the free market, was confidently forecasting that OPEC would collapse. The *Economist* magazine, which had predicted the glut at the height of the shortage, was now saying 'I-told-you-so' and pointing to the nonsense of high oil prices. And there were encouraging signs. Several oil-producers, including Kuwait, began offering long-term credit, as an equivalent to a discount. Abu Dhabi, the most phenomenal of the small Sheikhdoms, had been producing a million barrels a day in 1974, charging a high premium for its sulphur-free oil. By February 1975 the companies were refusing to pay the premium and production fell down and down to 500,000 barrels a day, until eventually Abu Dhabi were forced to agree to drop the premium. Libya, which was charging a higher price because it was closer to Europe, found that the companies refused to pay extra because freight charges were now much lower. Almost everywhere, the supplies were being cut; by April 1975 OPEC was producing 35 percent less than its total capacity.

But beyond these local adjustments, nothing happened. The basic price of oil, allowing for some discounts and credits, remained firmly where OPEC had fixed it. The big oil-producers, notably Saudi Arabia, Kuwait and Libya, simply cut back their production and kept their oil in the ground. The cartel was still holding fast.

How was it happening? The key was Saudi Arabia. As Yamani had predicted, many of the biggest producers did not need the extra money, and thus had no interest in undercutting the others with cheap oil in order to break them—as might happen with an ordinary commercial cartel. Saudi Arabia, far the biggest, was already beginning to play the role of Texas, prepared to cut back or step up production to keep the price steady.

But this balancing-process was being achieved painlessly, as gradually became clear, because the oil companies were really doing the countries' rationing for them. The members of OPEC had still not agreed on a pro-ration system, for the familiar reasons: they could not decide on the basis of restrictions. But the oil companies were already the masters of allocating supplies—as they had shown to the producers in the sixties and as they had shown to the consumers during the embargo. They were still performing their smooth global balancing-act; but now they were doing it, not to keep prices down, but to keep them up. Behind the blustering speeches of Middle East oil ministers about the evils of the sisters, they now privately admitted that they owed a great deal to them: 'Why abolish the oil companies,' Dr. Amouzegar said to me, 'when they can find the markets for us, and regulate them? We can just sit back and let them do it for us.'

The disturbing fact was slowly emerging that (in the blunt language of the business) the sisters were in bed with the producers. It was this co-habitation, or this 'indissoluble marriage', that had always been the aim of Zaki Yamani in putting forward his grand design of participation. As Walter Levy had warned two years before, the sisters had been lured into the tents of the producers, and away from the weaker charms of the consumers.

It looked for a time as if the producers, with their zeal after the embargo, might be moving too fast in their desire to take control

over their resources. Most of them, including Kuwait and Saudia
Arabia, appeared bent on 100 percent participation, or virtual
nationalisation, which was an obvious provocation of their new
partners. But they realised well enough that they must not go too
far, and even Abdulla Tariki, the old scourge of the sisters and
co-founder of OPEC, proclaimed himself in March 1975 in
favour of co-operation and joint ventures with the companies.
The producers stopped short of divorce; they made clear that
they would still allow the former owners of the concessions long-
term contracts to guarantee their supplies on preferential terms—
the security which was the object of any dependent. It was this
delicate calculation, of how to get control without ultimately
alienating the sisters, which partly underlay the long-drawn-out
negotiations in Saudi Arabia and Kuwait.

It was an ironic new turn to the long careers of the companies.
They had first been sent abroad in earlier decades, with the encouragement of their home governments, to do battle on behalf
of their consumers for cheap oil, bargaining relentlessly with the
producing governments. But they were now discovered to have
suffered the fate of so many other pioneers. They had gone native.
They were now in danger of becoming virtually the commercial
mercenaries of the foreign governments.

Aramco, proclaiming itself as the most successful American
corporation abroad, was now only American in the sense that it
was owned by United States shareholders. Its whole glittering
future—whether as oil producer, road-builder, farmer or universal
contractor—depended, as Frank Jungers made clear, on getting
still closer to the Saudi government. Aramco dare not now seek
any assistance from Washington. When they had tried to get State
Department support for the financial terms of participation, in
mid-1974, Yamani was so angry that (as he put it to me), he
decided to 'punish' them, by cutting down their profits, through
the agreement at Abu Dhabi in November 1974. The companies
backed down, and withdrew their objections; but the punishment
went ahead, and the partners were thereafter more conciliatory,
the more so since Yamani had learnt how to divide and rule them,
playing Mobil against Socal.

The ambiguity about Aramco's loyalties was again illustrated
when, in early 1975, the Pentagon was hinting that they might
have to invade oilfields in the Persian Gulf. Soon afterwards a
private Californian company called Vinnell was hired by the Saudi

Arabian government to assist in defence training, including the protection of oilfields. Visiting the Aramco compound soon afterwards, I was talking to a young Saudi engineer who expressed his bewilderment about the role of Vinnell: 'Are they supposed to be defending the oilfields *from* the Americans, or *for* the Americans?' In fact, Aramco was now in a similar position to the Vinnell employees; if there were to be an invasion, from whichever quarter, they would have to be on the side of the Saudis.

There was nothing consciously conspiratorial in the sisters' support for the OPEC cartel. The oil executives could honestly insist that they were constantly operating according to the market, trying to take oil where it was cheapest and refusing it where it was too expensive, as they had done in Abu Dhabi and Libya. It was the 'hidden hand' of the market, operating through the companies' computers, that was redirecting tankers from one loading port to another, not any dark scheme to maintain a cartel. Like Rockefeller's men at 26 Broadway a century earlier, the companies could well maintain that they were buying oil through the most efficient and economical system. But they were operating within a monopoly. The laws of OPEC allowed some variations in price, but ensured that it would not fall below a fixed level.

The companies had no real incentive to break the system, while they had many reasons not to. They were essentially passive instruments. Their dependence on the producers had been ordained by the participation agreements, by their long-term contracts, and by their own commercial rivalries. The members of OPEC knew well enough that to maintain their hold over the companies, they must keep them submissive and divided. 'The more confident we feel,' explained Yusuf Shirawa, the Oil Minister of Bahrain, 'the better for the oil industry . . . If I were running the oil companies, I'd have to be prepared to take all the blame, and absorb all the shocks between governments: if they try to divide, they will lose.' The companies were well aware of their need to be passive. They did take the blame, and absorbed the shocks, and they never dared publicly to complain about it; as one Shell director put it: 'That is our greatest difficulty—we have to be silent.'

The competition between the companies served not to bring down the price, but to increase the dependence of each company on each producing government, as they competed for long-term contracts and preferential terms. It was the paradox of the anti-trust muddle over the last two decades: the instrument that had been designed to enforce competition had ended by strengthening a foreign cartel.

The Western governments gradually came to realise that they could not look to the companies to break the cartel. In London BP privately told the Department of Energy that the company must not be expected to take a firm line against OPEC; BP depended too heavily on the goodwill of producers, and whatever the British government wanted to do, they must do themselves. In Washington the Church Committee, which published its report after some argument between Senators in January 1975, put little trust in the companies. The basic fact about oil, said the Report,[1] was that the big companies were no longer in a position to bargain with OPEC on behalf of the consumers. 'The primary concern of the established major oil companies is to maintain their world market shares and their favoured position of receiving oil from OPEC nations at costs slightly lower than other companies. To maintain this favoured status, the international companies help pro-ration the product on cutbacks among the OPEC members.' There were thus parallel interests between OPEC and the major oil companies, against the interest of the consumer: 'The companies ensure their access to the crude, but at the price imposed by OPEC, regardless of a theoretical crude oil surplus.'

The U.S. government (said the report) should therefore seek to cut the companies loose from this system of mutual dependence with the producers. The companies' joint purchase arrangements with the producers should be investigated by the anti-trust division, and the government should try to create uncertainty among the producers, as to their access to the consumer markets. But the report was not acted upon: the government lacked the political will to move against the companies, and the trust-busters, as we have seen, faced much greater difficulties once they were dealing with sovereign states.

THE RIGHT PRICE?

To the spectator of oil history, this new development might seem like the last act of a farce: all the characters were changing partners. But there was also another unexpected twist to the plot. For the Washington administration, which had been noisily battling to push down the price, taking up the fight that the companies had abandoned, was now itself increasingly muddled as to

what the price should be. While publicly insisting that the price
was outrageously high, the government was privately worried that
it might go too low.

For the United States, it was a culmination of the dilemma that
had begun in the 'forties, when the cheap Arab oil had first begun
to flow in: how to take advantage of this bargain energy, without
becoming dependent upon it? Even with the quadrupled price, the
imported oil was still cheaper than any likely new alternative for
the United States. The expectation of the free-market economists,
that the higher price—and the huge profits of the oil companies—
would stimulate enthusiastic new exploration inside America soon
proved illusory. In 1974 the oil companies were allowed to double
the price of 'new oil' from the United States, but the production
still fell during the year by 600,000 barrels a day. The prospect of
exploiting oil-shale from Colorado and elsewhere soon appeared
far more expensive than had been predicted. Nuclear energy was
still deadlocked with problems of safety, and Gulf and Shell, who
had an ambitious joint nuclear project in California, suffered large
losses. Other ambitious plans, from windmills to solar energy,
were likely to provide only a fraction of energy needs. These set-
backs suggested that, in the context of free trade and a free market,
the price of imported oil was not too expensive, but too cheap. And
the investors in energy, including the oil companies, now dreaded
that the Arabs could suddenly bring the price of oil right down
again, in the old Rockefeller tradition of price-wars, to undercut
potential rivals.

In February 1975 Dr. Kissinger came out with another brand-
new policy, formulated by his new energy expert at the State
Department, Thomas Enders, an outspoken giant of a man with
few inhibitions. Kissinger now proposed that there should be an
agreed 'floor price' below which the industrialised countries would
not allow oil to be sold, so that new investments and alternative
fuels would be safeguarded. The scheme had powerful logic behind it, but it was very difficult to decide where the floor should be.
While a floor of $6 to $7 seemed appropriate to protect coal or
domestic oil, a figure of $12 seemed necessary to protect such in-
novations as oil-shale: so that the floor seemed in danger of meeting the ceiling. And this came dangerously near to supporting the
Shah's own theory, that the price of oil must be kept high for the
sake of the West. In any case there was little chance of agreement
from most Europeans, who saw little prospect of developing serious

alternatives. The definition of the floor price was never fully worked out, and the idea was virtually shelved by March 1975.

In the meantime, President Ford had proclaimed in January 1975 a new policy; to establish tariffs on imported oil, rising from $1 a barrel as a means of cutting down consumption while encouraging internal production. But this policy, as Senator Church complained, marked a complete contradiction of previous statements. Having said that the price-rise was catastrophic, the administration now proposed to add a further price-rise, which gave OPEC a useful argument to raise prices still further. Church proposed instead a direct cutback of oil imports from OPEC of 2·5 million barrels a day, with gasoline rationing if necessary. Neither Ford's policy nor Church's was approved, and Americans remained as dependent as ever on OPEC oil. United States oil policy, as Anthony Lewis described it in the *New York Times*, had become 'as clear as sludge'.

Sheikh Yamani, in Saudi Arabia, professed himself to be totally confused by the contradictions of American policy. 'On the one hand,' he told Ray Vicker of the *Wall Street Journal*, 'the United States says oil prices are too high, and that it cannot afford the present level . . . Then you try to impose new taxes and measures to increase the price of oil on your domestic market. You contradict yourself.' Yamani was quite willing, he said, to co-operate 'when we can find out what that policy really is'.[2]

In Britain, the confusion about the price was rather more straightforward. For the next five years Britain, like the rest of Europe and Japan, would suffer a huge extra bill for imported oil. But after five years, the British expected to be largely self-sufficient from the North Sea, and by the spring of 1975 the Department of Energy was coming round to the view that Britain could and should become a net exporter (that is, should export more oil than she imported). America and Britain thus appeared to be exchanging their historical roles, with Britain becoming the exporter, America the importer. Clearly this implied that Britain would have an interest in maintaining or increasing the price; as one minister put it, 'We'll be propping up OPEC as hard as we can.' In Japan and the rest of Europe, except Norway, there was no such confusion: without alternative sources, they wanted the oil as cheap as possible. But the British and United States oil policies were now again moving in parallel, having turned round 180 degrees: both were determined not to let oil get too cheap.

While the public protests and even threats against the wicked OPEC cartel continued, many pressures were converging in favour of keeping high prices—which could only be achieved by keeping the cartel. The British and American governments, the seven sisters, the Texan oil lobby and the independent companies, the investors in alternative energy, the conservationists, and OPEC all agreed on the dangers of a price-reduction. Only the consumers disagreed—and the governments without hope of alternatives, most of all the countries of the Third World. It was a sudden and tragic new alignment in the endless battle between the haves and the have-nots: the have-nots, having staked their future on the cheap fuel, were confronted with a solid alliance to keep it expensive.

THE COMPANIES

Where, in this confusion of turnabouts, was the role or character of the sisters? Could the real companies stand up? At the beginning of the crisis the sisters appeared to be rapidly losing their world role. 'Actually they're paper tigers,' said Senator Church: 'they don't know what they own until they read it on the ticker-tape next morning.' They had, as they admitted, no leverage over prices; their concessions were rapidly being taken over or nationalised; and they were buffeted as never before between producers and consumers. While the producing countries were now able to secure their submission at one end, the consuming countries were determined to control them more closely at the other end.

The companies were being shorn of much of their profits from the oilfields 'upstream' and would thus have to look for more profits 'downstream', from the business of distributing and selling oil. It would be a painful adjustment, for the companies had been much less efficient at the selling end (except for Shell, who had always been better at selling oil than finding it). And the companies could only make more profits from consumers with the permission of the governments who controlled prices, which in turn gave home governments more potential control. Thus restricted, many oilmen in private depicted their future role as being similar to that of international utilities, like airlines, subject to constant oversight by national and international authorities.

But were they really so powerless? After all, the companies were still far the biggest global organisations in the world, with their great fleets, their pipelines, refineries and chains of filling-stations, which no country could hope to replace. More than ever, the great companies were now the archetypal giant multinationals, controlling huge interests across the world, not only in oil, but in petrochemicals and alternative energy, from coal to nuclear power. After the energy crisis the higher oil-price had pushed them still further up the league of giant companies. In *Fortune's* list of the biggest U.S. corporations, published in May 1975, Exxon had overtaken General Motors to be the world's biggest corporation, by sales: and the five American sisters were each among the seven biggest American companies, with only General Motors and Ford as their rivals.

With these great resources, why should the companies not again reassert their control of the market? If they gave up any share in the concessions, might they not become stronger rather than weaker? For with no special commitments to particular countries, they could divide and rule the producers without constraint, like other international operators, to beat down their prices. Rockefeller, after all, was able to establish his monopoly not by owning the sources of crude oil, but by controlling the markets without which the oil could not be sold. And the seven sisters, as any motorist can see, remain the kings of the market. The new national companies of the producers, like Petromin in Saudi Arabia or Sonatrach in Algeria, have not as yet dared to sell oil direct to the consumer.

But the size and scope of the great companies had now made them weaker rather than stronger in their battles with the Arabs. Their integrated structure, from selling to producing, made them more vulnerable to the concerted pressures from the producers, as access to preferential oil became more precious. The companies may yet divorce themselves from the concessions, voluntarily or involuntarily. They may be forced to by the consumer governments, or by a more militant mood in the producer countries, which may call for total nationalisation. But for the time being the companies seem trapped in participation, their eyes transfixed by their hunger for preferential oil. Behind all their American or British panoply, and their solemn statements about consumers and competition, their first interest lies in satisfying foreign producers.

It is difficult, as it has always been difficult, to see the big companies for what they are, to look through the emperor's clothes. They are so long-established and far-reaching, their executives so

well-paid and well-equipped, that the home governments are apt
to forget that they are simply trading companies with shifting
allegiances whose overriding aim is to make money. The mystique
of the great companies, which so successfully intimidated the
Sheikhs and the Shah, intimidated also the Western governments,
and many of their own employees. They have talked the language
of statesmen and sovereigns, encouraged by their long years of
dominance over the producers. Exxon proclaimed its dedication
to the American philosophy; BP has basked in its patriotic image.
But behind all these fine façades, they have been basically committees of engineers and accountants preoccupied like most
businessmen with profit margins, safeguarding investments and
avoiding taxation.

For some of the men at the top, the limitations are evident
enough: 'We were just tiny companies drilling for oil,' said one
Shell director, 'who have been simply caught up in world politics.'
But many company men were apt to be fooled by their own image,
easily assuming the arrogance of world rulers. And the home
governments, funking the problem, were content to let them lord
it. It might have been better for the West if the companies had all
been, like Texaco, straightforward buccaneers. If each had been
flying the skull-and-crossbones, the need for patrolling navies
would have been obvious. As it was, their high-sounding assumptions of global responsibility helped to deceive both the governments and themselves.

Could the companies, with more imagination and foresight,
have transformed themselves into quite different organisms, representing a far wider interest and thus able to avert or soften the
eventual crisis? They have had the limitations, of course, of any
private company responsible to shareholders in a few countries.
But the shareholders' interest is a long-term and global one, and
the sisters have been way ahead of other companies in their world
experience; Exxon, as they boasted, was a multinational corporation fifty years before the term was generally used. The big companies, as they revealed when they shared out the oil during the
embargo, had a unique machinery to enable them to be the world's
oil controllers; but they were basically so predominantly American
or British that other nations could never allow them that role over
the long run. If they had lived up to their rhetoric they might have
transformed their board and top management to reflect an awareness of both producing and consuming countries: thus they might

well have avoided the initial confrontation with OPEC, which was
born in the Exxon boardroom. As it was, Exxon like most of the
others remained ruled by the Texan engineers, with a sadly limited
political vision. (Dr. Kissinger, after one meeting with the oilmen
was heard to complain that they were a living disproof of the maxim
of Marx, that the captains of industry always know in the end
where their true political interest lies.)

Since the political uproar of 1973 the companies have made a
few gestures towards wider representation. Exxon in May 1974
elected their first woman to the board, Martha Peterson of Barnard
College. Gulf, under heavy fire for their large-scale bribery, took
the ingenious step in April 1975 of electing a nun to the board,
Sister Jane of Pittsburgh. BP admitted a trades unionist, Tom
Jackson of the Post Office Workers, as one of its two government
directors. But these innovations had little effect on the insulation
of the companies, and in the meantime there was little sign of any
restraint in the boardrooms' self-advancement in the midst of a
world recession: in 1974 all the American chief executives spectacularly pushed up their own salaries, with bonus payments up
by an average of 33 percent. The salary of the chairman of Exxon,
Ken Jamieson, went up to $677,000 a year, followed by the chair-
man of Mobil, Rawleigh Warner, with $596,000.

The major shareholders in some companies may eventually
insist on breaking up the self-perpetuating oligarchies of the oilmen. The bankers who control the largest blocks of shares are
more globally-minded and politically conscious than the engineers:
were they to intervene, they might be able to insist on new management and new objectives, to make the companies the visible representatives of the consumers' interests, which must be their only
long-term political justification. But for the present, the companies
are not seen to represent anything much more than their own
interests. In this situation, the Western governments will have an
obligation to impose far greater supervision over the companies,
for the delegation of the responsibility for oil is intolerable, so long
as the companies are not seen to represent the consumer. The
awareness that the companies are supporting the OPEC cartel will
be a continual source of indignation to the public; and even though
some Western governments may decide that the oil price must be
high, it will remain unacceptable that the price should be fixed in
Teheran or Riyadh.

The task will not be an easy one. The new state organisations, like the Federal Energy Agency in Washington or the British National Oil Corporation in Glasgow, may succeed in acquiring more effective information in the future; but the governments will have to make the companies accountable on a much wider basis, exposing them to congress or to parliament in a permanent dialogue. (Since the producers now know as much as they do, or more, about the business, much of the historical justification for secrecy is outdated.)

WHO CONTROLS?

The need both to supervise the companies and to widen their representation goes far beyond questions of foreign policy and commitments abroad. For the West's dependence on energy is now so fundamental and so expensive in terms of investment, that the planning and forecasting cannot be left primarily in the hands of private corporations. The decisions made in the companies' boardrooms affect not only the future of Abu Dhabi or Iran, but the whole character of the future Western life-style. According to the Chase Manhattan Bank's estimate in May 1975, the world oil industry will need $2·2 trillion ($2,200,000,000,000) between 1970 and 1985 for its optimal development, allowing for a growth of 4·5 percent a year in oil consumption: the best way to raise this huge capital, advises the Chase, is to allow banks to expand rapidly. But this assumption begs the question of whether the Western countries should be encouraged to become still more extravagantly dependent on oil. If the big companies are allowed to follow their projections, there is a real danger that the Western countries will become crippled by their indebtedness to this one seductive and fickle source of energy.

The failure of the Western governments to keep track of the companies and control them has marked their history since Rockefeller. It was never clear who was using whom. As the business became more global in the 'forties and 'fifties, so governments thought they were using their companies by encouraging them abroad, with anti-trust clearances, tax advantages and diplomatic support, while the companies were in fact far better at using them, in ways that were often against their governments' interest. Thus Harold Ickes, having safe-guarded Socal and Texaco in wartime, soon found them overcharging the U.S. Navy for oil. Thus the tax concession given to Aramco in 1951 became the instrument of the companies' growing dependence on the Middle East producers.

Thus BP, having been set up by Churchill as a more patriotic alternative to Shell, soon became equally uncontrollable, and landed the British government with the Abadan fiasco in 1951.

It was only in extreme crises that governments considered intervening, as Ickes did with the wartime Petroleum Reserves Corporation; or as the British proposed in the Anglo-American talks in 1953, when they suggested an International Petroleum Council, including both consumers and producers. But such a get-together did not suit the American companies or the State Department, and the discussions were abandoned when the oil again flowed freely and cheaply. For the subsequent evasions, the Western governments were at least as much to blame as the companies. The governments were content to let the companies operate their own policies, even while the cartels were confronting each other—a situation which an international council should have avoided. The more intelligent company men now knew that they were encouraging a 'fool's paradise' that could not continue, and that the cartel that was working on the side of the consumers could soon be turned quickly against them. A few, most notably in Shell, warned of the danger. But the governments did not care to involve themselves until they had to, and most oil companies preferred to keep the burden to themselves. After the final clash and the showdown in 1973, the governments were left to pick up the pieces, and to try to think out, for the first time in twenty years, a coherent oil policy.

But the formulation of a policy now depends on deciding who should be the controllers, and the problem of control is now a much more difficult one, for it involves agreement between nations. That oil must be controlled by some kind of organisation appears to be indicated by its whole astonishing history, when it has so consistently flowed into the hands of a monopoly or a cartel. But the present system of control has the worst of two worlds. The two rival cartels with which this book begins and ends, of the companies and the countries, are equally unsatisfactory and partial. They basically have represented two extremes of Anglo-American consumers, and of Arab producers. Even though OPEC may maintain its hold on its members for some time, it is a fragile alliance which will be acutely vulnerable to political pressures all round it. However much the members of OPEC may seek to buy off the discontent of their neighbours or the third world, the sole control of such a concentration of wealth and power will be a perpetual provocation to needier countries.

To establish any acceptable international control system has been made much harder by the polarisation of the past history. If the oil companies had evolved in the 'sixties into more genuinely international companies they might have helped to serve as a bridge between the two sides. As it was, they first confronted the producers as a solid cartel, then provoked the producers cartel, and then virtually joined it. The attempts by the consuming countries to form a new counter-cartel, through the International Energy Agency, seem likely only to help OPEC to unite by forming a common bogey: 'The IEA is the real cartel of sovereign states,' said Ali Atiga of OAPEC: 'they want to fix the posted price in Brussels or Paris. The trouble is, that the West and the Arabs have now got a mutual complex about each other.' 'If you insist on a confrontation,' said Yamani, 'you'll stop us from helping you.'

The conjunction of the price-rise with the embargo made the problem of establishing a fair system much more difficult. Without the embargo, the price would certainly have gone up, but less precipitately, allowing a gradual adjustment to the increase. As it was, the embargo, which constituted an intolerable use of oil as a political weapon, became confused with the quite separate question of prices, making the confrontation with OPEC more emotional and bitter.

But the interests of OPEC and the consuming countries are likely to come closer together. On the Western side, as we have seen, there are important elements now in favour of the high oil-price, and governments and bankers are seeking to connect up the OPEC surplus funds with the industrial economy. On the OPEC side, the producers realise that their stability, and the security of their investments, are dependent on the industrialised countries who alone can safeguard them. The producers cannot afford to live too long in a hostile world; and they must seek to give their cartel the legitimacy of acceptance by the rest of the world.

They are likely to find it easier to be accepted by the industrialised countries than by the countries whose economics are much more seriously threatened, the countries of the Third World. There will never be fairness in the world of oil, for in no commodity has the contrast in distribution been more extreme, between the haves and the have-nots. But a more equitable means of control may emerge, not through the clash of two cartels, but through the establishment of some kind of international council, such as was put forward twenty years ago, which could begin to produce some kind of synthesis between the aspirations of the chief parties in the triangle—the producers, Western consumers and the Third World. It is possible that OPEC, which does not need all its new money, may prove a more effective bridge between the rich and poor countries than all the earlier aid-giving institutions of the West. And without the middle-men of the oil companies the West will have to face more urgently the problem of sharing resources.

The only moral justification for the existence of OPEC, as its members (led by Algeria) have realised, is to present themselves as the representatives of the rest of the Third World. And though they may play this role with a good deal of humbug and false piety, it may prove a useful performance. For as OPEC becomes more deeply involved in questions of food and other commodities, so the first harsh purpose of the cartel is likely to be softened. Both producers and consumers will be trying to enlist the support of the Third World, in a new version of the earlier battles for the uncommitted world.

The long bitter battles for the control of oil, with which this book has been concerned, have not been without their compensating side. For as the first injustices of the Standard Oil monopoly played a part in forming the national consciousness of the United States, and building up Federal controls, so the subsequent conflicts have played a role in arousing a global awareness. And though oil companies have contributed to some very disagreeable episodes in history, from the support of dictators to the bludgeoning Arab embargo against Israel, they have also helped to bring the world closer together and, partly through their sheer excesses, have compelled governments to interest themselves in wider perspectives.

In his apologia for the first Rockefeller, Allan Nevins likened him, together with the other 'heroic' industrialists in nineteenth-century America, to the Elizabethan sea-captains, like Drake, Hawkins and Cabot, who roamed the world in search of their booty. The comparison is apt, and the subsequent companies, operating across the globe out of sight of their governments, still retained some of the independence of the old adventures (like Drake with Queen Elizabeth) working sometimes on behalf of their home governments, often in defiance of them. The governments have gradually tried to catch up with the runaway industry. But in the meantime the mysterious fluid which appeared in Pennsylvania has become the life-blood of each industrial country, and the companies which purvey it have become the very centrepiece of the economic system. It is an appropriate symbol of this dependence that one of the present Rockefeller family should now be the head of one of the biggest banks, the Chase Manhattan, and another the Vice-President of the United States.

From the beginning the extraordinary substance created unique new problems and new kinds of organisation. Its erratic production encouraged cartels and centralised control. Its unequal distribution, its fluid state and its dependence on transport gave it a special ability to bring closer, first the United States, then the rest of the world. The pipelines and tankers that carried it became the arteries of the industrial world, but in doing so rapidly changed the political balances. For a time the old 'Coal Powers' which dominated the nineteenth-century world seemed able to dominate the new 'Oil Powers' through technology and cunning: but the new balance was bound to follow, as the oil powers discovered their strength. Often enough the effects of the first exploitation of oil, whether in Mexico, Iraq or Venezuela, were dangerously disruptive, leaving a legacy of bitterness that still jeopardises relationships. But oil can be a solvent as well as an explosive ingredient, once it has been integrated into the political system of nations, as it helped to unify the United States, or to reconcile Iran and Iraq. Whether destructively or constructively, oil has encouraged extremes.

It would be quite wrong, I believe, for the Western nations to regard the change in the balance of oil power as a sudden new external threat to the essential Western civilisation. For it is clear from the history of oil that much of the opposition to the companies, and the formation of OPEC, was the result of liberal tendencies within the United States and Europe, as much as the emergent nationalism in the Middle East and Latin America. The 1952 Report by the Federal Trade Commission was the textbook for Arab technocrats to understand the Oil Cartel: the Labour Government's nationalisations in Britian were the cue for Dr. Mossadeq to nationalise Iranian oil, and thus increase the Middle East militancy. The anti-trust movement, which was both magnificent and muddled, effectively prevented long-term coordination between the companies and Washington. The Western governments had decided—without perhaps fully realising it—

that they were not prepared to sanction a thorough-going cartel or to use force if necessary, as the means to ensure cheap oil supplies, and the brutal tactics of the 'twenties and 'thirties were no longer acceptable. Thus OPEC, from the beginning, was armed with ammunition from liberals and trustbusters, and the Shah could count on some sympathy from Western public opinion, when he attacked the wastefulness of the Western way of life, or the connivance of the sisters. The suddenness of the transition, when it came in 1973, took the West completely by surprise; but the seeds of the change were already present in the Libyan episode of 1970, when it was clear that the big companies could not 'black' nationalised oil as they had done in Iran two decades earlier.

OPEC thus owes its success in large measure to the internal divisions in the West, and to the democratic instincts behind the anti-trust movement. There is always a great difference, in oil as in labour negotiations, between advocating the sharing of power, and having power wrested away: but the suddenness of the confrontation and the transition was partly due to the West's inability to face up to a new situation until forced to. Having faced it, the West has now, like any far-sighted management, to reach an understanding with the producers on their many points of common interest. But the danger is that, as in union agreements, it will be the consumers and the outsiders that will suffer; the object of the Western liberal conscience must be no longer OPEC but the under-dogs of the Third World.

The Western governments can no longer evade the problem of control of the companies, which they have evaded so often in the past. Politicians and diplomats have to face the full implications of the fuel that has helped to build up their world, and to relate the operations of the companies to systems of democratic choice and control. The public must know the full cost, both economic and political, of their dependence on the energy which they have come to take for granted. It would be absurd to present this as a simple question of Congressional enquiries and parliamentary debates, or of high-minded statesmen confronting greedy tycoons. It is a long and difficult process of making a bridge between a highly technical global industry and the ordinary lives of consumers. But this task is essential, if the Western public is not to find itself lured into a technological trap, becoming caught up in an extravagant life-style which they will not be able to afford, either economically or politically: and no purely profit-orientated industry can be expected to portray objectively the full implications.

NOTES

1. Multinational Report: January 2, 1975.
2. *International Herald Tribune*, March 20, 1975.

16

POSTSCRIPT:

The New Crisis

For four years after 1974, the world's oil supplies seemed to have reached some kind of uneasy stability. OPEC continued to meet, year after year, to put up the oil-price; but the Saudis continued to play a moderating role, as they had promised, and the increases did not keep pace with world inflation. The Western nations still encouraged the image of OPEC as an Arab cartel which had succeeded in holding them to ransom; but the reality was really very different. For the OPEC system, supported by the Seven Sisters, had provided a mechanism in many ways more secure than the companies' cartel which preceded it, since it rested on the assent of the producing countries. The old saying about the Sisters could now almost be adapted to: 'if OPEC did not exist, it would be necessary to invent it'.

The producing countries' demands were modified partly by the fears of Western inflation and disruption from the moderate members, and partly by the awareness that there was a margin of surplus oil-production, some of which came from outside the club. In North America the Alaskan oil supplies, coming down through the pipeline to the United States, had helped to ease the worries about a shortage; and by 1978 Mexico was proclaiming vast new oil discoveries by their state oil company Pemex which, according to some experts, were on a scale approaching the Saudi reserves. It now seemed possible that the United States, having devoted so much diplomacy, arms-dealing and compromising to maintain its influence with the Saudis and Iranians, might find its energy salvation in closer relations with its southern neighbour, whose earlier oil reserves the West had previously ruthlessly exploited.

The most spectacular new source of Western oil was in the North Sea where Britain was able, partly as a result of the higher oil-price, to develop the oilfields so that by the end of 1978 they were providing three-quarters of her home consumption. The cost of this self-sufficiency, both financial and political, was very high: sixty per cent of the North Sea oil was controlled by foreign companies, most of them American; and the generous tax concessions devised to encourage exploration resulted in an unexpectedly low tax revenue from the British government: BP had made an operating profit of over two billion dollars from the single Forties Field in 1978, but the tax yield amounted to only eighteen percent. The state oil company, the British National Oil Corporation, had acquired participation in the North Sea fields but still had great difficulty in discovering the true profits of the companies, or even the destination of the oil; as its chairman, Lord Kearton, complained in 1979, 'the oil companies don't like to have an informed government'.

Moreover the economic benefits to Britain of this new bonanza proved in many ways illusory: the security of the oil enabled the pound sterling to improve its value against the world's currencies, but the very strength of the pound was a major obstacle to Britain's industrial exports, and there was little sign of the oil profits being used to provide the industrial revival on which Britain's long-term future depended. It was as if the old curse of oil, which has so often ruined other nations, was now working its spell once again. But in spite of all these dangers, Britain had the temporary advantage of being, alone among major industrial nations, the only one with the prospect of self-sufficiency in the most valuable fuel. And the North Sea oil was a valuable addition to the world's energy balance.

But in the United States, which was by far the biggest factor in the energy equation, the imbalance was becoming still more dangerous. The higher oil-price had failed to produce the incentives for domestic oil exploration and developing alternative fuels; the consumption of gasoline continued as extravagantly as ever; and even though industrial companies had achieved some saving, the dependence on imported oil continued to increase. The arrival of President Carter and his Energy Secretary, James Schlesinger, promised major initiatives to reduce consumption and to encourage domestic energy; they proclaimed a new policy, to allow higher prices together with higher taxation of companies, explaining that the energy challenge was the 'moral equivalent of war'. But the new policy soon came up against the full weight of opposition of both Congress and the oil companies. Congress was resistent to any policy which would produce higher prices for the consumers and increase oil company profits, and they were doubtful of their ability to impose a compensating windfall tax on company profits. The oil companies were publicly opposed to any windfall tax even though privately they accepted the need for it and the smaller Independents (whose political clout in Washington was stronger than the giants') lobbied ruthlessly to oppose any such tax.

The cheapness of American oil continued to be an obstacle to any rational conservation policy, and the price of gasoline at the pumps continued to be half that in Europe, which led to increasing transatlantic friction. But Congress ruled out the imposition of a direct tax on gasoline as politically unacceptable; and the public distrust and hatred of the companies got in the way of a more long-term solution.

The Seven Sisters and the other big companies in the meantime did little to diminish that distrust. They still insisted that they needed higher profits to finance furthur exploration and development of alternative fuels, but their operations did not inspire much confidence. Since the 1973 oil crisis they were investing less in oil operations abroad, and they also had diminishing confidence in America's own energy prospects; so they began to move more heavily into alternative industries. Gulf moved more heavily into chemicals, Occidental moved into coal, Atlantic Richfield bought the second biggest American copper company, Anaconda. Mobil extended its ancillary activities, and was even found to be growing tomatoes in Britain.

Exxon bought into electronics, culminating in a billion dollar bid in 1979 to buy the Reliance electrical company, on the excuse that it was developing energy-saving motors. All the Seven Sisters in the meantime stepped up their publicity and patronage to make themselves politically more acceptable; and Mobil, proclaiming itself defiantly as the 'Big Mouth' of the oil industry, became the Maecenas of oil, financing television series, exhibitions and theatre while continuing to lecture the public, both in America and Europe, about the indispensability of the oil giants. All the oil companies insisted that only they could undertake the necessary investment to solve the world's energy shortage; and when in 1978 the World Bank initiated a modest plan to finance exploration on the Third World, Exxon protested indignantly to the Treasury

Secretary Michael Blumenthal about this trespassing on their territory. Yet behind their public confidence, there were signs that the Sisters were beginning to lose faith in the future of their fuel.

The Seven Sisters remained the dominant corporate powers in the United States and in 1978 they provided four of the six biggest companies (by sales) in the Fortune Directory. The movement to break up the giants grew stronger after the 1973 crisis, and both the Senate and the House of Representatives initiated hearings. Many Senators supported divestiture, whether 'horizontal' requiring the oil companies to sell off their non-oil investments) or 'vertical' (breaking up the big companies into smaller units responsible for different sectors—exploration, pipelines, refining or marketing).

There is little evidence, I believe, to suggest that divestiture would in itself increase the supplies of American domestic oil, and the rhetoric of the divestiture campaign conceded a good deal of populist deception. But I believe there remain strong arguments for vertical devestiture (as I testified to the Senate in February 1976): firstly because the size and power of the giants has served to obstruct rational policies inside the United States, producing a deadlock which can only be broken by anti-trust action; and secondly because the links between the big companies and the OPEC producers have provided contractual relationships creating a new kind of joint cartel, with no real incentive to compete with lower prices and with a dangerous confusion of purpose. I believe that this close relationship with OPEC, the continuation of Sheikh Yamani's 'indissoluble marriage', helped to conceal the true facts of the continuing energy crisis. The companies were continuing to earn easy profits from their long-term contracts, but the security of that supply was very uncertain, and a sudden cut-back of two or three million barrels a day could create a world shortage.

It was the fall of the Shah at the end of 1978 that precipitated the new oil crisis: as so often in oil history it was an unexpected political event—like the Suez war in 1956 or the Arab-Israel war in 1973—which transformed the world market and created a new context. The revolution in Iran was full of diplomatic ironies. The Shah had in the end become the victim of his own financial success; the huge new revenues that came from his quadrupling of the oil price in 1973 had helped to generate a web of corruption, inflation and dissatisfaction which led to his downfall. And the five years of

OPEC unity which had begun with the Shah'a victory in December 1973 ended with his departure from Iran.

But even after the exile of the Shah and the closing down of the Iranian oilfields the danger was slowly to register in the capitals of the West: partly because of Western oil stocks, but partly because of the faith in Saudi Arabia. The four Aramco partners had for long expected that the Saudis could, when necessary, increase their production from the regular 8.5 million barrels a day to perhaps as much as 12 million; but when the crisis came the Saudis stayed at around 8.5 million. Partly it was because of technical difficulties, which had been suspected but indignantly denied by the companies, but partly because the Saudis had discreetly changed their own internal policy. They had become exasperated by the Western wastage of oil, and the continued demands to produce more; the internal pressures for conservation were becoming stronger; and they had decided that spare capacity could no longer be the means of maintaining their leadership of OPEC. They kept down their production, and in the long cold winter of 1978–79 the oil companies ran down their stocks, without Western consumers realising it. Even when the Iranian oilfields were re-opened they did not make up the difference.

The change in Saudi policy, together with the unpredictable new regime in Iran, were at the heart of the new world oil crisis. But neither were part of any OPEC conspiracy on prices. In fact the Saudis resolutely kept their price down, and the first sign of rising prices early in the new year came not from the Arabs but from the British oilfields in the North Sea, which being closer to the market, with short-term contracts, reacted more rapidly than OPEC; and the oil companies as usual found the opportunity of higher profits irresistible.

At the same time there was frenzied activity on the Rotterdam 'spot' market which rapidly achieved a special importance as the indicator of the crisis and the extreme price-levels; and this market for short-term deals on the margins became the wild frontier of the oil-trade. The market is called Rotterdam because it is in that huge harbour that oil can quickly be bought from the tankers, and Dutch brokers have specialised in quick bargains. But the 'spot' market can operate wherever there are banks, big money and clever men with an international telephone and the right contacts. Today the most important centre is probably Geneva, where oil traders can buy and sell oil with the support of the Swiss banks who can provide loans in total discretion, and the bribes and pay-offs through secret bank accounts which are essential for adventurous oil deals.

As the world shortage became more serious in early 1979 other factors combined to make the spot market more excitable. Two countries, South Africa and Israel, were desperately anxious to buy oil, for they had been boycotted by all the Arab producers and by the new Iranian regime; and they were both prepared to pay above the OPEC prices. And in the meantime the new Iranian regime decided to sell more and more oil on a day-to-day basis, instead of on long-term contracts, so that they in effect became part of the spot market. The telephones between Teheran and Geneva became hectic, and at Kargh island, where Iranian oil was taken, the situation was increasingly tense: the tankers waiting for oil on contract remained anchored, while tankers buying the expensive oil at spot prices went to the head of the line.

Through the cold spring of 1979 the oil shortage began to be felt across the industrial world, and by the week beginning May 14 the spot market first began to go crazy. On the Monday the price of Iranian oil was $23 a barrel—already far above the normal OPEC price. On Tuesday it rushed up to $28. By the end of the week it had hit $34. The first reaction to these exorbitant demands from Iran was that no one would buy; but it was soon clear to brokers that there were very big buyers. One deal was made for 2 million barrels at $34 a barrel which turned out to be on behalf of Texaco; and Mobil and Shell at Rotterdam were both known to be buying at over $30. The huge sums could only be paid by the Seven Sisters, or other big companies, and it was evident that they were playing a large role in bidding up the Iranian oil. As one broker in Geneva put it to me: 'The Iranians and the Seven Sisters understand each other very well'. The basis of their understanding was simple; they were both happy to see the oil-price go up. And at the same time the companies were selling cargoes of their own cheap oil bought at the long-term contract prices, making huge profits. For the Saudis, who were trying to keep prices down, the situation was becoming humiliating.

The spot market became still more excited as the Europeans tried to replenish their diminishing oil stocks: the Italians had only about two weeks stocks in their tanks, and the Spaniards and Swedes were in similar difficulties. The high prices began attracting oil from other parts of the world, and American companies

with refined oil in the Caribbean began sending it to Rotterdam
instead of the East Coast of America. Washington, already acutely
worried by the American shortage, suddenly offered a subsidy of
$5 a barrel for heating oil, to lure the oil from the Caribbean back
to the States. The subsidy did not (according to Geneva sources)
stop American oil coming to Europe; but it helped to bid up the
Rotterdam price.

At this point the spot market became the focus of a major
diplomatic row. The Europeans were now furious at a subsidy
which competed with them without (they complained) doing
anything to mitigate the real problem — the wastage and lack of
exploration within the United States. The rift between the two
sides of the Atlantic on energy policy had never been wider. But
the Europeans were divided, as so often on energy problems before.
The French wanted all European countries to abstain from high
bidding, but other countries were too desperate to agree. The
Germans offered the British industrial investment in return for a
share of the North Sea oil, but the British were increasingly
determined to keep hold of their own oil supplies. Only the
Japanese had successfully cut back their demand since 1973,
reducing their oil imports by as much as 120 million barrels a
year. The scramble for oil had set all nations against each other, as
it had six years before. But the Rotterdam market had revealed
that this was a different kind of crisis from the 1973 crisis. It was
based on a more serious shortage, after the cut-back in Iran, and it
was no longer a straight-forward battle between the countries of
OPEC and the industrial nations. For OPEC, as well as for the
West, the situation was dangerously out of control. One Geneva
broker told me: 'here, we think OPEC has nothing to do with it'.

By mid-June the oil shortage was becoming still more acute. In
the United States the lines of cars waiting for gasoline, with all the
extra wastage involved, were the outward sign of a wider panic
than in 1974, while the questions of price-levels or rationing were
still unresolved. Nor could the Europeans reach any sensible
agreement: the competition for the marginal oil was not simply
between the companies; behind them, in many cases, the national
governments were negotiating and pushing, conscious that they
would be blamed for any shortage at the tanks.

In this tense atmosphere, the members of OPEC assembled at
the end of June for a critical meeting to settle the oil-price,
appropriately enough in Geneva. While the Swiss oil-traders were

still making their quick bargains for spot oil, from their bleak little offices in downtown Geneva, the Arabs, Iranians and Latin Americans were gathering in the Intercontinental hotel up the hill, surrounded by Swiss police and intensive security precautions, to try to work out an agreed price that could stabilise the system.

The scene at the Intercontinental hotel had much in common with the OPEC meeting in Vienna four and a half years ago: once again the world's press had flown in, conscious that the decision would affect every country's economy. Once again Sheikh Yamani was the star of the show, immaculately bearded and dressed, thoroughly attuned to his international repercussions. But the awareness of the economic and diplomatic importance of oil now went much deeper, since the recession of 1974–75 which had been partly caused by it. The two meanings of the word power—in both the physical and the political sense—had come much closer. The critical importance of energy was underlined at the simultaneous meeting in Tokyo of the seven Western leaders from the United States, Canada, Japan, Britain, France, West Germany and Italy.

At the same time the instability of the oil system was much more evident, and the arrival of the unknown delegation from the new Iranian regime was a reminder to all the others of the fragility of the oil powers. Sheikh Yamani was visibly more worried than he had been four years earlier, and with good reason; for the Saudis were in a tense predicament. On the one hand the Iranian revolution had made them much more aware of their isolation as a conservative, monarchic nation: they had become much less confident of their relationship with the United States, which had paid them in a declining currency, which had failed to curb their oil consumption, which had been unable to protect the Shah, and which had encouraged an agreement between Egypt and Israel to which the Saudis were opposed. On the other hand, Yamani and most of his government still felt very responsible for the world's economic stability; and the Saudis were desperately anxious to eliminate the frantic spot trading which the Iranians had ruthlessly escalated. The old rivalry between the Saudis and Iranians which had raged through the sixties was now made more tense by their opposite political philosophies, at two extremes of Islam.

After long and bitter negotiations the OPEC nations eventually announced an awkward compromise with a two-tier system for the pricing of oil; the ceiling price would be $23.50, but the Saudis would only raise their official price to $18 in an attempt to restrain the increases: 'we did our best, but I can't say I'm really happy', Yamani explained at the Intercontinental hotel: 'probably this high level will gradually come down. That is my guess. It all depends on you consumers'. The Saudis were very aware that they were no longer in a position either to make or break OPEC, for they could only increase their production by around a million barrels: they could no longer play the kind of equalising role that Texas had once played inside the United States. And the OPEC delegates realised that the effectiveness of their organisation was much more uncertain, after the maverick role of Iran and the chaos on the spot market. 'The law of supply and demand will now decide the price of oil', said Yamani, 'Countries like Britain and Norway are for most of the time ahead of us in prices'.

But the Western leaders in Tokyo, meeting for a summit conference and, hearing the news from Geneva, preferred to put all the blame on OPEC for what they called an 'unwarranted rises in oil prices ... that will lead to more unemployment, more balance of payment difficulties and will endanger stability in developing and developed countries alike'.

They promptly pledged themselves to limit their oil imports to the levels of 1978, and to enforce much stricter conservation: 'We are going to have to change our habits', said President Giscard d'Estaing of France: 'invent a lot, improvise a lot'. Certainly the effects of the price increases would be grim for the world economy. The new prices, which added about 15 percent to the cost of oil to the West, would add between 1.5 and 2 percent to the rate of inflation, causing higher unemployment and lowering the standard of living. For the countries of the Third World, who were most hit by the increases, the bill for oil imports would go up by $9 billion to $27 billion dollars; and although OPEC agreed at Geneva to allocate 800 million dollarrs to the special fund for developing countries, this could not compensate for the disruption of Third World economies.

But in the light of the market chaos that had preceded the OPEC meeting, it was naive to blame OPEC alone for the higher prices. It was the failure of the industrial powers to tax and restrain imports which had been the cause of the crisis, as much as the Iranian revolution and its consequences. The instability that resulted from the shortage was almost as worrying to the Saudis as to the industrial nations. It had brought up all the old problems about the control of oil and of the giant companies, which had

POSTSCRIPT: THE NEW CRISIS

once again benefitted from the crisis with quick bargains and higher long-term profits. The communique from the Tokyo summit did imply that the Seven Sisters had played some role in the pushing up of spot prices: 'we agree to take steps to bring into the open the working of oil markets by setting up a register of international oil transactions. We will urge oil companies and oil-exporting countries to moderate spot market transactions.' And they recognised by implication that the cheap American oil was distorting the world system: 'we agree on the importance of keeping domestic oil prices at world market prices or raising them to this level as soon as possible'. The seven powers in Tokyo were attempting to enforce what was in effect the beginning of a consumers' cartel, to counter the power of OPEC. But they did not really hold out much prospect, either of effective control of the oil companies, or of lasting international agreements between consuming governments.

The crisis had revealed once again the deep ambiguity in the characters of the oil companies. On the one hand they were virtually global utilities, with their networks of oil-rigs, tankers, refineries and filling-stations. On the other hand they were shrewd and ruthless day-to-day traders, constantly watching down the line for where profits could be maximised and shortages exploited. The Western world was back again with the problem: 'who controls ?'. And they faced a growing nightmare of uncontrolled corporate power: that companies could lead the public still further into a helpless dependence from which they could not rescue them; and while they blamed governments, they could frustrate them from any intervention. In the difficult years ahead all the Western nations are likely to be faced with having to move away from the oil-based, car-based, air-conditioned economy on which they have been built. But while the oil companies maintain their corporate clout, whether in America or Europe, they will inevitably stand against any such drastic adjustment; and they will continue to promote their chief product while offering no security for its future or cheapness. The bold decisions—to reduce consumption through taxes, to switch to alternative energy sources, to begin changing the pattern of transport—can only be taken by governments and voters.

But the problem of controlling the companies is only part of the broader question of national self-discipline and restraint. It is convenient for the Western powers to use OPEC as their scape-

goat, and to criticise the Middle Eastern nations for their reluctance to increase production. But the insatiable consumption of
the West is putting those nations, particularly Saudi Arabia, in an
impossible position with regard to their internal and external
politics; and the only effective escape from a future oil crisis will
not lie with increased Middle Eastern production, but with more
serious Western conservation and economies. The real lesson of
the crisis can only be: we have met the enemy—it is *us*.

Index

Index

Compiled by Robert Urwin

The use of italics, e.g. 50(*18*), *59* – means on page 50, note 18 refers to notes – page 59.

ALSO BY ANTHONY SAMPSON

THE SOVEREIGN STATE

For the first time the secret workings of a multinational conglomerate are revealed in devastating detail.

'An original, topical, sensational, truly significant book'
Evening Standard

'A lucidly written and very readable account of ITT. Mr Sampson's book is likely to be widely read'
Sunday Telegraph

'A very good book'
J. K. Galbraith in *The Sunday Times*

'Reads with the pace of a thriller'
Yorkshire Post

'Mr Sampson cuts through the words with directness and a responsible readability. His book will reach a wider audience than any comparable study so far'
The Times

CORONET BOOKS

ANTHONY SAMPSON

THE ARMS BAZAAR

Anthony Sampson's riveting new book plunges the reader into the clandestine world of the billion dollar arms trade. Beginning with giants like Krupp and Vickers, Sampson analyses the Lockheed and Northrop scandals and the job of the contemporary arms merchant. With the escalating arms trade, Middle East deadlock and nuclear proliferation, this powerful and impressive book gives a real insight into the workings of the world's most global industry.

'Brilliant – has the impact of a major scoop'
New York Times Book Review

'Personalities and human drama, international affairs, relationships and rivalries within and between huge companies and governments are woven into this gripping story'
Daily Telegraph

CORONET BOOKS

ALSO AVAILABLE FROM CORONET BOOKS

ANTHONY SAMPSON

All these books are available at your local bookshop or newsagent, or can be ordered direct from the publisher. Just tick the titles you want and fill in the form below.

Prices and availability subject to change without notice.

CORONET BOOKS, P.O. Box 11, Falmouth, Cornwall.

Please send cheque or postal order, and allow the following for postage and packing:

U.K. – 55p for one book, plus 22p for the second book, and 14p for each additional book ordered up to a £1.75 maximum.

B.F.P.O. and EIRE – 55p for the first book, 22p for the second book, and 14p per copy for the next 7 books, 8p per book thereafter.

OTHER OVERSEAS CUSTOMERS – £1.00 for the first book, plus 25p per copy for each additional book.

Name ....

Address ....

....